Enterprise Innovation

Series Editor
Jean-Paul Bourrières

Enterprise Innovation

From Creativity to Engineering

Michele Missikoff
Massimo Canducci
Neil Maiden

iSTE

WILEY

First published 2015 in Great Britain and the United States by ISTE Ltd and John Wiley & Sons, Inc.

ISTE Ltd
27-37 St George's Road
London SW19 4EU
UK

www.iste.co.uk

John Wiley & Sons, Inc.
111 River Street
Hoboken, NJ 07030
USA

www.wiley.com

Library of Congress Control Number: 2015944024

British Library Cataloguing-in-Publication Data
A CIP record for this book is available from the British Library
ISBN 978-1-84821-851-2

Contents

Preface

This book is organised into four parts and contains the chapters written by the partners of the BIVEE Project that provide an elaborated view of the project rationale and the achieved results.

The partners of the European BIVEE Project are:

– Engineering Ingegneria Informatica S.p.A, Italy (Project Coordination);

– Università Politecnica delle Marche, Italy (Scientific Coordination);

– AIDIMA, Instituto Tecnológico del Mueble, Madera, Embalaje y Afines, Spain;

– ATOS, Spain S.A.;

– BIBA – Bremer Instituts für Produktion und Logistik GmbH, Germany;

– CNR – National Research Council, Italy;

– BOC Asset Management GmbH, Austria;

– Loccioni Group, Italy;

– SRDC – Software Research, Development and Consultancy Ltd.

The consortium also enjoyed the cooperation of City University London in the area of business creativity.

The BIVEE consortium wishes to thank the European Commission, DG Connect, Unit A3, the Project Officer and the reviewers for their continuous attention and the constructive observations that allowed us to constantly improve our work. This has led to a very positive conclusion of the project that has been granted with the Excellence Award.

We hope that this book will be interesting and stimulating, both for researchers and developers, suggesting to them new directions to improve the technological solutions for innovators, and for practitioners, who will be exposed to new opportunities and solutions to engage in the challenging adventure of innovation with a better perspective of success.

We wish to acknowledge the Easy Chair platform for the available portals and services that made the editing of this volume possible.

Michele MISSIKOFF
Massimo CANDUCCI
Neil MAIDEN
June 2015

PART 1

BIVEE Project Framing

1

Business Innovation in Virtual Enterprise Environments

Virtual enterprises are temporary alliances of business organizations that come together sharing skills, knowledge, information and resources in order to better respond to business opportunities, and whose cooperation is supported by information technology (IT) platforms. This chapter is an introduction to business innovation in virtual enterprise environments.

1.1. Introduction

One of the most interesting aphorisms in innovation literature is, without doubt, "Change Or Die!". Few people know that this sentence was actually an advertisement made by Electronic Design and was conceived in 1970; only in 2005 did it become the title, and the opening concept, of Alan Deutschman's article "Change or Die" for the magazine Fast Company. In this article, talking about the contents of the previous IBM's Global Innovation Outlook conference, Deutschman argued that science has shown that only one time out of nine, when faced with preventable conditions like heart attacks, are people able to change the environmental conditions and the style of their life.

Moving the point of view from people to companies, we realize that the situation is exactly the same: companies, too often, are very slow to react to changes in environmental conditions and, when the environment is related to their own market, the results could be fatal. History is full of situations in

Chapter written by Massimo CANDUCCI.

which large companies have failed to change, although all the signs of a change in their target market were there, and consequently died.

So, the first thing, from a company's point of view when the market is going to change, is a willingness to change. In fact, as we know, willingness is a very important trait, but of course it is not enough. A company needs something else and this element is often what makes the difference between surviving and dying; this element is innovation.

In the literature, there are several definitions for the term "innovation", my favorite being: "innovation is the process of creating values from ideas" [TID 14].

The first concept we have to understand is that *innovation is a process*. It is not a spark that enlightens the dark minds of some business managers or engineers when they are closed in their office staring at the white ceiling. It is a process, with a beginning, with phases (or better "waves" as we will see later in this book), with owners, with stakeholders, with positive people and negative ones, with some organizational entities prepared to pay and others prepared to enjoy the success of a new product, a new service or a new way to do things in a better way. All of these actors work in a single body: the organization.

The innovation goal is not only, as it might seem at a superficial analysis, to earn more money; that is only a side effect and, actually, is related only to product or service innovation from a company point of view. In a more general vision, we can have different levels of innovation, from a company level to a system and social level.

From a company point of view, we can have an innovation based on new and advanced materials which can be used to improve the product and service performances. However, we can have an innovation based on a new kind of product (or a new market for existing ones) or the need to update the existing products in order to make them more attractive for the evolving market. In this case, of course, the main innovation goal is related to the company's money.

From the society point of view, however, we can have a radical innovation based on new general interest findings or inventions, such as steam power, ICT revolution and bio-technology. In this case, we can find new technologies or products that, regardless of their brand, could change the way we do some daily activities. For example, we can listen to music using old vinyl records, usually only using very large devices, or we can use new portable devices with tens of hours of music inside and, in the near future, other kinds of portable devices with all the world's music directly available through streaming. When these types of innovation are more related to improving the quality of life of people, for instance a new green engine, we can say that the social impact of innovation is much more important for society than it is for the company that puts it on to the market.

1.2. Business innovation and virtual enterprises

Innovation is not an art, craft or skill, it is not an empirical practice or an engineering activity and it is not a creative vision. In fact, it is a manifold compound that includes all these ingredients and one more: knowledge. Knowledge is the real key factor that makes the difference between disorganized and theoretical innovation attempts and real business innovation strategies. People involved in innovation processes are well informed about company business strategies, products and services, the technologies involved, new market opportunities and other relevant technologies or methodologies recently discovered or just placed on the market.

Unfortunately, it is very difficult to have all of this knowledge in one team. Even in very large, smart and agile organizations, the covering of every aspect of the innovation process is a mirage or a myth. This is the reason why typical one-brand organizations (like companies) need to link themselves to more complex structures, in order to share their knowledge with other partners in a particular context and, on the other hand, to access other partners' knowledge, building a so-called virtual enterprise.

A virtual enterprise is a temporary alliance of business organizations that come together sharing skills, knowledge, information and resources in order to better respond to business opportunities, and whose cooperation is supported by IT platforms.

In this larger vision, related to virtual enterprises, business innovation could be described as a designed, managed transformation of some aspects of the enterprises involved aimed at a substantial improvement of the quality of products and services. This results in better production processes, cost reductions (and/or revenue increases), better staff satisfaction and better overall sustainability.

In this extended and more open and fertile environment, we will find many opportunities in which we can insert new ideas and insights directly coming from the virtual enterprise's knowledge.

First, we will discover new products or services for our usual market and, similarly, new potential markets for our current products and services catalog. This new range of prospective opportunities could be considered enough to say that the virtual enterprise has great potential, but this is only the beginning.

Sharing knowledge with partners also means the possibility of improving business production, and administrative processes; this is a very important topic because having more efficient processes directly means having fewer costs and often better quality.

At the same time, the company will have access to competencies, capabilities and skills not usually present into the employee headcount, this is a crucial aspect when in a situation a particular, and not easy to find, skill is requested.

Last but not least, the company will have access to technology directly offered by technological partners, with the best economic conditions and in a partnership model that, as everybody knows, is much more efficient than the classical customer–supplier relationship.

In such a supportive environment, there are great advantages not only for standard production projects, but also especially for innovation projects that can have great opportunities precisely because they were developed in a knowledge-based environment where every partner has access to information, the best practices, tools and platforms that allow the best selection of ideas in the described scenario and generate the best conditions for great innovation projects.

1.3. Bibliography

[TID 14] TIDD J., BESSANT J., *Strategic Innovation Management*, Wiley, 2014.

2

From Creativity to Innovation: the Importance of Design

This chapter highlights the importance of creativity to innovation, and outlines the role of design as the discipline and the activity by which to deliver innovations from the results of creative thinking. It summarizes advances in design thinking principles and processes, and draws on the author's own experiences with creative design processes to make some simple recommendations about the use of creativity techniques and design artifacts to enable more effective design innovation.

2.1. Creativity and innovation

The recent years have seen a growing interest in the adoption of creative and innovation processes in enterprises. Creativity and innovation are perceived to be increasingly important means by which enterprises can distinguish themselves in marketplaces. The strategic importance of creativity has been acknowledged by many commentators, both at the international level – the Nomura Research Institute's proposition is that creativity is the next economic activity, replacing the current focus on information – and within the UK, where the Cox review commissioned by Gordon Brown in 2005 saw exploitation of the nation's creative skills as "*vital to the UK's long-term economic success*" [COX 05]. The Cox review concluded that: "The success of the creative industries notwithstanding, there is evidence that

---

Chapter written by Neil MAIDEN.

UK business is not realizing the full potential of applying creativity more widely". Other European nations have similarly identified the importance of creativity and innovation to their macroeconomic growth.

Having said that, although many organizations are seeking to adopt processes to support and adapt creativity and innovation processes, the differences and relationships between these phenomena are, on the whole, poorly understood. This chapter argues that, if we are to develop new processes, methods and tools to improve creativity in industrial innovation, these organizations need to better understand the relationship between creativity and innovation. Therefore, the chapter explores the role of design to deliver the results of creative thinking through to innovation in enterprise settings more effectively, and introduces a set of creativity techniques and design artifacts that can be embedded into service design work in order to support creative thinking more effectively.

2.2. Creative problem-solving methods

Creativity has been the subject of research in different disciplines for much longer than research in business and enterprise. The Greek philosophers considered the nature of creativity in human endeavor. In the late 19th Century, mathematician Henri Poincaré reflected on the nature of successful creative thinking in science [HAD 54]. Creativity research in its current form was developed during World War II in the U.S. military, and grew in the 1950s when Osborn and Parnes [OBS 53] and Synectics [GOR 60] developed new creative-solving processes. During the 1960s and 1970s, leaders such as Edward De Bono developed lateral thinking [DEB 07] and Genrich Altshuller evolved the theory of the resolution of invention-related tasks (TRIZ) method [ALT 99] for structured creative problem solving (CPS). More recent creativity research has been undertaken in disciplines including cognitive psychology [CSI 96], artificial intelligence [RIT 01] and product design [MOG 07]. The result, today, is a large and multidisciplinary body of knowledge of theories and models, and large collections of processes, techniques and tools for CPS. Many researchers would agree with the following prototypical definition of creativity:

> *the ability to produce work that is both novel (i.e. original, unexpected) and appropriate (i.e. useful, adaptive to task constraints)* [STE 99].

Most of the use of creative processes and techniques to deliver new and useful outcomes in business and enterprise today is based on the CPS methods that were developed from the 1950s onward. The first of these was the CPS method of Osborn and Parnes [OBS 53]. This method was originally intended to help people understand and use their creative talent more effectively. It supports six stages of problem solving: objective finding, fact finding, problem finding, idea finding, solution finding and acceptance finding. The six stages are arranged into three groups – understanding the problem, idea generation and planning for action. Multiple versions of the method have been developed over the last half-century, and it has been successfully applied to resolve many problems creatively in business, enterprise and other domains.

Central to this CPS method is an open exploration or search for ideas to generate many novel and varied ideas and new perspectives, and then focus thinking by identifying ideas with interesting or exciting potential to refine, develop and put to use. The method primarily uses traditional forms of brainstorming and related creativity techniques in this divergent phase of CPS. The application of the CPS method to support creative thinking about, for example, the redesign of an urban bicycle hire scheme, leads to the generation of a large number of ideas and concepts, normally documented on post-it notes, which are sorted and prioritized. Although useful, a method to support the development of a large and complex service such as cycle hire needs other processes and artifacts, both to support divergent thinking about complex systems and to support convergent thinking to form and validate complex concepts, processes and systems.

2.3. Linking creativity and innovation through design

If creativity is the ability to produce work that is both novel and appropriate, innovation is generally accepted to be the process of translating these ideas into goods or services that generate value or for which customers will pay, or as Steve Jobs once called it, the creativity that ships. More recently, however, this view of innovation as translation from idea to good or service has been challenged, especially in the design world.

For example, the UK Design Council has defined design as the process that shapes creative ideas to become practical and attractive propositions for users – creativity deployed to a specific end [DES 11].

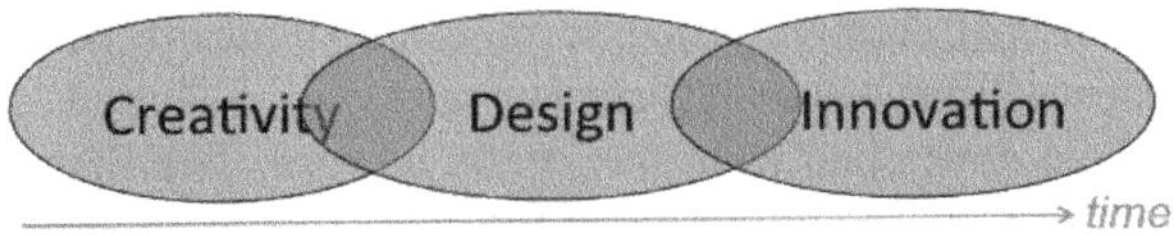

Figure 2.1. *The relationship between creativity, design and innovation*

The obvious interpretation of this definition is that design represents a critical intersection between creativity and innovation – that without design, creative ideas cannot be shaped to become the practical and attractive propositions that users need and want. These relationships are shown in Figure 2.1.

Evidence from reviews of many successful products and services tends to support this view. Many of the products and services that we consume and most people would consider to be creative are the result of substantial design processes – the iPhone, the Dyson vacuum cleaner and the FIAT 500 automobile are all creative and innovative, but all have been designed to the nth degree. Indeed, it is difficult to think of many of today's creative products and services that have not been designed in some form.

Therefore, one conclusion to draw from the Design Council's definition of design, creativity and innovation is that enterprise solutions that are to innovate must be designed, with creative ideas as inputs to the design process. Increasingly, business and enterprise solutions must be subject to design processes.

2.4. Service design processes

Service design thinking is a human-centered innovation process that involves observation, collaboration, fast learning, the visualization of ideas, rapid prototyping and concurrent business analysis [LOC 10]. Although design thinking was originally developed to be applied to the design of consumer products such as iPods, shavers, lamps and chairs, and services such as hotel check-in and post office counters, it can also be applied to a wider range of design projects such as business processes for new staff appointments and unloading aircraft activities, and wider socio-technical solutions for air traffic control and awarding parking tickets.

Different definitions of service design have been offered, and most considered it to be an interdisciplinary activity. For example, Moritz in

[STI 10] reported that *it helps to innovate (create new) or improve (existing) services to make them more useful, usable, desirable for clients and efficient as well as effective for organizations ... a new holistic, multi-disciplinary, integrative field.* Livework, also in [STI 10], described service design as *the application of established design process and skills to the development of services... a creative and practical way to improve existing services and innovate new ones*. However, one of the most incisive definitions is from the service design studio 31 Volts in [STI 10], which describes it as when *you have two coffee shops ... selling the same coffee at the same price, service design is what makes you walk into one and not the other.*

Moreover, Stickdorn and Schneider [STI 10] identified five principles of service design. These principles are:

1) *It is user-centered*: services should be experienced through customer's eyes using techniques such as observations and walkthroughs of a service from a customer's perspective.

2) *It is co-creative*: all stakeholders should be included in the creative process, and empowered to generate ideas, contribute to the designs and undertake creative work themselves. This principle assumes that all stakeholders have knowledge and experiences that can contribute to the design process.

3) *It sequences*: services should be visualized as a sequence of interrelated actions using representations that are described later. The essence of a service is a sequence of activities – activities that take place before the service used, and end after the use of the service is complete.

4) *It is holistic*: the entire environment of service should be considered. The wider context of the service – its physical, business and cultural environment – influences and is impacted by the design of the service, and needs to be considered.

5) *It evidences*: intangible services should be visualized in terms of physical artifacts to enable them to be reasoned and communicated about.

Central to evidencing of a service is the notion of a touchpoint. Service consumers will interact with a service through a sequence of touchpoints. Moreover, the design of a service's touchpoints and the relationships between these touchpoints provide the basic elements of service design represented with design artifacts, as is described later in this chapter.

Most of the services are implemented through touchpoints, which are different elements of a service that are used by its consumers. The touchpoints of a service should be designed to be harmonious with each other, and to deliver a more complete service experience to its consumers. Moreover, the touchpoints should convey the brand story to be communicated through the service. As such, service touchpoints can take multiple forms, and are not just the traditional interactive systems that we associate with a service. For example, touchpoints of the service that is offered by the London Underground include not only the trains, ticket machines and gates but also the staff uniforms, the fonts of the station names and the world-famous logo. Two examples of such touchpoints are shown in Figure 2.2.

Figure 2.2. *Examples of touchpoints from the London Underground*

In our experiences with creative service design processes, design specifications such as floor plans, wiring diagrams and floor layout specifications, and design artifacts such as sketches, storyboards and paper mock-ups have been critical to the success of these processes. Indeed, design artifacts fulfill a range of important roles in service design, and differentiate design from established CPS processes. For example, these design artifacts enable stakeholders to express their knowledge about the structure of services and touchpoints, in order to facilitate a longer, more continuous creative service design process. These artifacts also provide the service designers to externalize knowledge about the new service design and the creative ideas about it, to communicate this knowledge to others, and then to enable the validation, reflection and learning of this knowledge more effectively. And, of course, design artifacts enable design knowledge to be stored for reuse later and/or in other service design projects.

To this end, service design methods and techniques offer a range of established types of design artifact. Based on [STI 10], these include:

– *Stakeholder maps*: visual or physical representation of actors that are involved in service expressed using simple diagrams. Each map can be used to describe stakeholder's interests and motivations, and links and relationships between these actors and stakeholders presented in visually engaging ways.

– *Expectation maps*: these chart user expectations of interactions with an existing service expressed visually to capture and describe these stakeholder's emotions using media coverage. The maps can be used to capture emerging consumer higher level expectations. Expectation maps are often used for service diagnoses in order to discover areas needing attention.

– *Desktop walkthroughs*: these walkthroughs take place in small-scale three-dimensional (3D) models of the service environment built out of props, toys and other objects such as Lego bricks. These simple-to-use environments are used to act out service scenarios in the environment in order to explore emerging interactions. In particular, their physical nature encourages creative collaboration and play.

– *Service role-play*: service role-plays are interactive experiences that help stakeholders contribute to improving a service experience. Stakeholders enact different situations involving customer contact with the service. Different stakeholders fulfill different roles in turn to enact outline ideas generated using other creativity and design techniques. Role-playing service scenarios are particularly effective when observed by others as a source of new idea generation. Service role-plays are particularly effective for designing touchpoints and improving empathy with customers.

– *Customer journey maps*: a customer journey is a structured visualization of service user experiences that describe interactions with different touchpoints. Each visual map is a concrete description of touchpoints. The content of a map can include a customer's own content to facilitate empathetic engagement, and analysis of them can be used to identify problem areas and opportunities.

– *Storyboards*: storyboards are graphic organizers – illustrations or images displayed in sequence to previsualize a motion graphic or interactive media sequence. Storyboards are frequently used to develop designs in film, theater, business and interactive media. Most present a sequence of events in visual form, and this accessible form helps to involve service consumers in the service design. Because storyboards describe the wider environment of a service, they contextualize the design of the service and each touchpoint.

– *Design scenarios*: a design scenario depicts one or more hypothetical stories, each of which details a service offering. A design scenario can be expressed in different forms, from plain text and sketches to videos, and any of these representations can be used to specify service touchpoints and interactions, and can be refined with personas representing the service consumers. Moreover, each scenario can be manipulated during walkthroughs as a positive or a negative scenario, for example how could things be made worse?

– *Service prototypes*: service prototypes enable the simulation of future services in order to facilitate learning by doing. The use of service prototypes exploits a behavior that is common to service consumers – to only be able to express what they want when they see or recognize it. As such, service prototypes are great for engaging consumers in the emerging service design. Service prototypes can take different forms, from informal role-play conversations to detailed full-scale creations with physical touchpoints. Our experience is to build prototypes for specific purposes.

Examples of some of these types of artifact applied to the redesign of an urban cycle hire scheme are shown in Figure 2.3. The left-hand image in Figure 2.3 shows a desktop walkthrough of urban cycle use in different city park settings and the right-hand image shows one storyboard of the future service design use.

Figure 2.3. *Examples of two service design artifacts applied to the redesign of an urban cycle hire scheme – on the left a desktop walkthrough and on the right a storyboard*

In the author's experiences, the development and use of these different types of artifact are pivotal to service design processes, and provide an important connection between earlier creative thinking and later innovation processes. For example, service prototypes from the design process can continue to be developed, deployed and refined continually through to innovation processes – translating the ideas embodied in each service or design prototype into goods or services that generate value or for which customers will pay. Returning to service design thinking as a human-centered innovation process that involves observation, collaboration, fast learning, the visualization of ideas, rapid prototyping and concurrent business analysis [LOC 10], we observe that there is less focus on creative thinking, which we perceive as a weakness that is discussed in the next section.

2.5. Integrating creativity support more effectively into service design methods

Service design's early focus on observation, collaboration, idea visualization and rapid prototyping contradicts some established early creative thinking activities to collect, relate and, more importantly, to incubate. Indeed, the focus on the rapid development of visualizations and prototypes can impede incubation leading to transformational creativity that transforms the nature of the design problem and, hence, the service to be developed. Therefore, a challenge is to integrate more effectively creativity techniques into established service design processes.

One solution to overcome this challenge, which we have been experimenting within creative service design projects, is to couple selected creativity techniques with different types of design artifact. The coupling seeks to recommend design artifacts to use to both capture the results of creative thinking with the creativity technique, and to support further convergent creative thinking through the artifact as part of the design process. This coupling of technique-with-artifact is determined by an analysis of the mutual affordances of the technique and artifact. In the remainder of this chapter, these couplings are demonstrated by three pairings of creativity technique and design artifact that have been applied successfully in our design projects:

1) constraint removal with design sketching;

2) creativity triggers with desktop walkthroughs;

3) idea combination rules with storyboards.

Each is described in turn.

Constraint removal is a transformational creativity technique in which a problem solver brainstorms possible constraints on a problem, and then removes one or more constraints in turn to generate a new, larger space in which new ideas are then generated. Constraints are descriptions that limit the space of possible ideas, and Onarheim [ONA 12] reports that constraints both constrain and foster creativity. Design domains are often overconstrained, for example in the design of office buildings, therefore most of the time constraints need to be relaxed. In contrast, many artistic domains often lack constraints, for example in poetry, thus artists often impose constraints to add artistic value to their work. There are four possible strategies for creative constraint handling: (1) black-boxing constraints, which treat constraints as unchangeable in order to focus effort on other constraints; (2) removing constraints, which ignores one or more constraints in order to trigger new constraints, whether this is possible or not; (3) revising constraints, which reinterprets one or more constraints expressed in a design brief and (4) introducing self-set constraints, in order to structure and focus into creative design process. After an open brainstorm to generate possible constraints, relevant constraints and the handling strategies are selected, and a brainstorm takes place in the enlarged ideas space generated for each constraint in order to create a wide range of new ideas that increase design ideas. We couple constraint removal with design sketching, a simple and flexible technique to depict and evolve the new ideas generated from the constraint removal. Design sketches are effective in early creative work because it is rough and quick, allowing new design ideas to evolve quickly. It allows the creative team to focus on the concepts rather than on fine-grained details of the design, which can be addressed later. It is inclusive and accessible, as most people can sketch with a minimum of training. And it is fun, encouraging play and the right types of emotion for creative thinking. Therefore, a design sketch captures and evolves ideas from constraint removal more effectively than simple post-it ideas, and each sketch brings a design artifact that can be shared in the creative design process.

Creativity triggers are an exploratory creativity technique that guides creative problem solvers to discover and generate new ideas with certain predefined qualities in an existing search space. Each trigger defines a simple rule with guidelines, and is often derived from expertise in problem domains that reveals common qualities or characteristics of creative ideas that are generated during convergence processes. Two creativity triggers are *participation* and *convenience*. The *participation* trigger is derived from observations that people want to participate more in processes and businesses, so the trigger directs the creative process to discover new ideas that result in greater participation in the new design. The *convenience* trigger is derived from observations that people want either convenience, or at least the perception of convenience. Hence, the trigger directs the creative process to either remove one or more steps from a stakeholder's process, or to undertake one or more steps on behalf of the stakeholder. The typical outputs from the application of creativity triggers are more concrete, focused ideas that have some direct application to a design problem. Therefore, we couple creativity triggers with desktop walkthroughs, the artifact that takes place in small-scale 3D models of the service environment built out of props, toys and other objects. These simple-to-use environments are used to act out service scenarios in the environment in order to explore emerging interactions based on the concrete design ideas often generated from creativity triggers.

Finally, an idea combination rule is an instance of a combinational creativity technique that provides explicit rules that, when applied, find new ideas by combining two or more ideas already found in the space. It assumes that a search space has already been examined in order to discover new ideas, but further ideas can be generated from informed combinations of these ideas. Idea combination rules direct the creative problem solver to discover these new, informed idea combinations. Examples of these rules include combine two unconnected touchpoints, deliver known service with new types of touchpoint, apply known service to new information, and combine two actors into a single role. We combine idea combination rules with the storyboard design artifact to support more effective combinational creativity. Storyboards organize illustrations or images that are displayed in sequence to previsualize a motion graphic or interactive media sequence – these structures provide an effective mechanism by which to report, combine and share the results of combinational creativity.

Moreover, studies of how designers used design artefacts during creative design projects revealed other affordances and benefits of shared artefacts in early creative design processes. For example, in the EU-funded COLLAGE project[1], studies of design behavior revealed strong evidence of social flow during the production of shared artifacts in convergent creative processes to achieve a clear goal. This observed social flow with artifact use contrasted with the relative lack of evidence of flow during the earlier divergent search activities, during which individual participants often worked separately, in different ways, with different tools, to achieve different goals, and the participants did not demonstrate much evidence of social flow. Furthermore,

1 http://projectcollage.eu.

the empirical evidence revealed that individual and group creative thinking, which generated new knowledge during periods of social flow, happened at different times in the process to the learning about this new knowledge. Moreover, the shared artifacts fulfilled an important role in representing knowledge so that it could be learned. We observed cycles of:

1) social flow, in which participants in the creative process collaborated to generate new knowledge, often in the form of a shared artifact;

2) then, once the social flow had ceased, longer periods of reflective learning in which the same participants reviewed the shared artifact, sought to make sense of it, critiqued it, and learned new knowledge from it.

This observed relationship between creative thinking and reflective learning is not new. For example, the EU-funded FP7 Mirror Project developed an integrated model of creative thinking and reflective learning that was applied to individual and social creativity in different sectors[2]. Results from the COLLAGE project are consistent with this model; therefore, we propose to extend its model of social creativity and learning with elements of the Mirror descriptive model of computer-supported reflective learning. The model was developed from existing models of reflective learning (e.g. [SCH 83, BOU 85] and [LAV 93]) to be a reference model with which to develop computerized support for individuals, teams and organizations. Not all of the discrete activities described in the model need to be present in all instances of reflective learning. For example, during reflection-in-action during work, there might not be a conscious decision to reflect, and the framing of reflection might be brief. Furthermore, although the activities are expected to take place in a broad sequence, some activities can be concurrent and repeated.

2.6 Conclusions

This chapter has reported on the important role of creativity to innovation, and highlighted the role of design as the discipline and the activity by which to deliver innovations from the results of creative thinking. It has summarized recent advances in design thinking principles and processes, and draws on this chapter author's own experiences with creative design processes to make recommendations about the use of creativity techniques and design artifacts to enable more effective design innovation. It adopts the UK Design Council

---

2 http://www.mirror-project.eu.

definition of design as the process that shapes creative ideas to become practical and attractive propositions for users – that design is creativity deployed to a specific end [DES 11]. The implication is that enterprises that seek to innovate need to ensure that not only do these enterprises have effective processes, climates and organizational structures to support creative thinking, but also these organizations need to have adopted design processes sufficient to deliver innovations from the outcomes of these creative processes. The emergence of the design thinking and service design disciplines provides a set of processes, techniques and tools to deliver potential innovations from the results of creative thinking, but only if the enterprise's creative and design thinking processes are sufficiently integrated.

2.7. Bibliography

[ALT 99] ALTSHULLER G., The Innovation Algorithm: TRIZ, Systematic Innovation, and Technical Creativity, Technical Innovation Center, Worcester, MA, 1999.

[BOU 85] BOUD D., KEOGH R., WALKER D., *Promoting Reflection Experience into Learning*, Routledge Falmer, pp. 18–40, 1985.

[COX 05] COX G., Review of Creativity in Business: Building on the UK's Strengths, HM Treasury, 2005.

[CSI 96] CSIKSZENTMIHALYI M., *Creativity: Flow and the Psychology of Discovery and Invention*, Harper Collins, 1996.

[DEB 07] DE BONO E., *How to Have Creative Ideas*, Vermilion, 2007.

[DES 11] DESIGN COUNCIL, Design for innovation: facts, figures and practical plans for growth, available at http://www.designcouncil.org.uk/, 2011.

[GOR 60] GORDON W.J.J., *Synectics*, Harper and Row, New York, 1960.

[HAD 54] HADAMARD J., *The Psychology of Invention in the Mathematical Field*, Dover, 1954.

[LAV 93] LAVE J., CHAIKLIN S., (eds), *The Practice of Learning. Understanding Practice: Perspectives on Activity and Context*, Cambridge University Press, Cambridge, p. 20, 1993.

[LOC 10] LOCKWOOD T., *Design Thinking*, Allworth Press, New York, 2010.

[MOG 07] MOGGRIDGE B., *Designing Interactions*, MIT Press, 2007.

[OBS 53] OBSORN A.F., *Applied Imagination: Principles and Procedures of Creative Problem Solving*, Charles Scribener's Sons, New York, 1953.

[ONA 12] ONARHEIM B., Creativity under constraints: creativity as balancing 'Constrainedness', PhD Thesis, Copenhagen Business School, December 2012.

[RIT 01] RITCHIE G., "Assessing creativity", *Proceedings AISB-01 Symposium AI and Creativity in Arts and Science*, The Society for the Study of Artificial Intelligence and Simulation of Behaviour, pp. 3–11, 2001.

[SCH 83] SCHON D., *The Reflective Practitioner: How Professionals Think in Action*, Temple Smith, London, 1983.

[STE 99] STERNBERG R.J. (ed.), *Handbook of Creativity*, Cambridge University Press, New York, 1999.

[STI 10] STICKDORN M., SCHNEIDER J., *This is Service Design Thinking*, BIS Publishers, 2010.

3

The BIVEE Project: an Overview of Methodology and Tools

The EU needs an effective exit strategy from crisis, with special attention being paid to small and medium-sized enterprises (SMEs) that represent 99% of the enterprise active in the European production system. To this end, innovation appears to be a key factor in relaunching the EU industrial system. The business innovation in virtual enterprise environment (BIVEE) project proceeded for almost four years to develop a rich framework, i.e. a methodology and a cloud-based software environment, that includes business principles, models, and best practices, plus a number of advanced software services, to support and promote production improvement and business innovation in virtual enterprise (VE) environments (essentially, enterprise networks.)

3.1. Framing

The European industrial system needs to change pace to relaunch its competitiveness, after a long downturn. To this end, new strategies need to include renewed attention to continuous production improvement and, in a seamless way, business innovation. Both are essential for business survival in highly competitive markets where it is increasingly difficult to differentiate between products and services. Rethinking the European value production system in a way capable of integrating different ideas, people and cultures, is the only way for Europe to compete with the impetuous progression of the BRIC (the rampant Brazil, Russia, India, China) economies.

Chapter written by Michele MISSIKOFF and Pierluigi ASSOGNA.

Innovation is a term that represents a complex domain, requiring both domain expertise and a large amount of knowledge. First, domain knowledge (on the specific industry sector), but also knowledge on technology, business models, finances, markets, etc. Carrying out an innovation project is not an easy job for a single enterprise; it is really a challenge for a network of enterprises, having the problem of achieving the necessary coordination and synergy in a distributed, multipolar decisional and operational structure. We will refer to such networked enterprise structure as a VE. The objective of the BIVEE project has been to build methodological and technological solutions to improve the value creation capability of European SMEs, leveraging on constant production improvement and continuous business innovation.

In conceiving the project, we were aware that there are a number of barriers that EU SMEs still face and which prevent them from systematically adopting the emerging paradigms of collaborative business innovation. Among these barriers, we may cite the high fragmentation of the SMEs industrial fabric, the inadequate organizational and technological culture, the limited resources (such as financial, management expertise and competencies) available for SMEs, the lack of systematic connections with research institutions, and often their naïve and unfocused approach to innovation. However, there are a number of new business models and ICT solutions for supporting and fostering innovation in SMEs, such as Open Innovation, crowdsourcing, new forms of social networking, that we decided to acquire and transform into usable solutions for the European SMEs.

In BIVEE, we assumed a broad holistic approach to business innovation, tackling the methods, concepts, ideas, inventions and artifacts aimed at renovating in a sustainable manner the way enterprises conduct their production processes, considering that all is connected, at the business, technical, economical or social level. This is in line with the policy of the European Commission (see the Innovation Union Flagship initiative[1]), and in particular the strategy aimed at portraying the next generation of innovation-driven enterprises: only a synergic combination of new technologies, new business models capable of supporting and respecting our social and political foundations, could succeed in making the European enterprise system evolving and pushing forward the entire socioeconomic system.

Innovation largely remains a people-centric, "brain intensive" activity. From this perspective, BIVEE proceeded in building a distributed,

1 http://ec.europa.eu/research/innovation-union/index_en.cfm.

collaborative, knowledge-intensive framework, where not only emerging ICT solutions but also innovative business models, novel management methods and cooperative working styles have been integrated to the benefit of interoperable VEs.

The methodological framework of BIVEE is based on two interlaced spaces: value production and business innovation space. These are the concrete spaces, where resources and people cooperate to achieve the business goals, but at the same time they correspond to two knowledge spaces where the reality, inside and outside of the enterprise, is extensively represented in digital form. According to the BIVEE approach, such a rich knowledge base is used in the value production space (VPS) activities to support the monitoring and improvement of highly distributed production processes, and in the business innovation space (BIS) activities, to support the continuous innovation in a VE.

On the technological side, the logical architecture of the BIVEE software environment has been structured according to the two mentioned virtual spaces (VPS and BIS). Each of which is managed by a software platform: mission control room (MCR) and virtual innovation factory (VIF), respectively. Furthermore, to manage the VE knowledge supporting the activities of the two mentioned spaces, a platform for the management of a production and innovation knowledge repository (PIKR) has been developed. The three platforms stand on top of the Raw Data Management and Service (RDM&S) systems that interface with the local (to the individual SMEs) information systems, providing the necessary data interoperability and service integration for such diverse information sources. Among the offered services, we may list the support to a rich gamut of business activities: from production monitoring to market watch, from innovative think-tank support to technology watch, appraisal and evaluation of production and innovation activities, until risk assessment and prevention. An important position is given to innovation monitoring based on a number of key performance indicators (KPIs) capable of tracking the progress of an innovation project and, subsequently, assessing the effects of the adopted improvement and innovation solutions. The four main platforms have been based on the Open Software Architecture paradigm and advanced meshup techniques, by extensively reusing resources available in the FLOSS world and leading edge solutions provided by BIVEE partners.

To validate the BIVEE Environment, the project addressed two trial cases: the first trial concerned the area of distributed co-design, having innovation as a core focus (in Loccioni – General Impianti); the second trial took place on a traditional production/logistic chain (AIDIMA). The involved industry sectors have been carefully selected to validate BIVEE in very different cases: in fact, the latter concerns the "mature" wood and furniture industry, while the former is positioned in the hi-tech sector of robotic and automatic measurement equipments. To achieve a thorough assessment of the BIVEE Framework, the trial cases have been organized into two main phases: Phase 1, the First Monitoring Campaign (FMC), where the activities of the two VEs have been monitored in their "as is" practices, assessing their production and innovation performances before introducing the BIVEE platform; then, after the introduction of BIVEE, in Phase 2, we carried out the Second Monitoring Campaign (SMC), appraising the production and innovation performances once the BIVEE solutions have been adopted. Confronting the results collected in the two monitoring campaign has been a very useful activities that yielded to important indications for the implementation of the final version of the BIVEE Framework, released at the end of the project.

3.2. The mission of BIVEE

The mission of BIVEE is to provide advanced solutions to boost productivity and innovation capabilities of networked SMEs. To this end, BIVEE developed an integrated framework, i.e. a coordinated set of business methods, enterprise models and ICT solutions, to support adaptive, distributed, interoperable business ecosystems (in our case also referred to as Virtual Enterprise Environments: VEE) in pursuing continuous, distributed optimization and innovation practices.

Production optimization and innovation need to be addressed in a synergic fashion. About the former, the main focus of BIVEE is on manufacturing that is mainly based on well defined, deterministic value production processes where tangible and intangible assets are managed in order to achieve some business objectives, in a cost effective way. Here automatic equipments, such as robots, measuring sensors, various actuators, play nowadays a very important role. The BIVEE solutions concern the services aimed at monitoring the distributed production activities and intervene to correct deviations to the planned programs and/or to introduce all possible improvements. In these cases, the interventions do not structurally modify the enterprise organization or production maps (e.g. according to the *Kaizen* approach).

Innovation concerns the capacity to conceive and introduce a marked discontinuity, such as the introduction of a new product on the market or a new production process, which will impact on various enterprise dimensions, such as human resources, production means and methods, organization and finances, marketing strategies, etc. Innovation is a human-centric activity, often non-deterministic or serendipitous, in most cases starting with imprecise objectives, presenting a number of criticalities in the planning, forecasting and scheduling the activities, and in managing the risk. Here BIVEE proposes a specific set of services.

The lifecycle of an innovation project has three basic phases: the *inception*, the *evolution* and the *conclusion*. The inception typically starts according to five (not orthogonal) basic approaches: (i) *push-mode and technology driven*, when the innovation is generated on the supply side (e.g. by the appearance of a new technology); (ii) *pull-mode and demand driven*, when the innovation is motivated by a user need (i.e. by the future adopters); (iii) *co-creation*, when all the stakeholders cooperate together to generate product or process innovation (i.e., open, collaborative approach); (iv) *endogenous*, when ideas come from within the enterprise of the ecosystem; (v) *exogenous*, when ideas come from the rest of the world. We believe that the case (iii) is the most effective inception, although the hardest to achieve. In fact, innovative co-creation requires that different people, with different cultures, skills and needs (and, sometimes, conflicting objectives) share a cooperation space spanning across the whole product lifecycle, from R&D, to design, to production, until after-sales, and across the enterprise boundaries. When stakeholders belong to different realities (different enterprises, but also different roles, different geographical areas, including also external research organization), often remotely interacting, with different levels of engagement and timing, then the added value is more important. A situation that is typical of Internet-based social networking.

The central phase of an innovation project is where ideas are consolidated and progressively elaborated, pushing them forward until an applicable innovative solution is achieved. This is the phase that will be elaborated in detail in this book, with a characterizing point of view, concerning innovation projects carried out in a distributed, collaborative open organizations' network. Then the book addresses also the final phase of innovation and its interactions with the production space; in case of successful innovation, the outcome will be transferred to production and, eventually, to the market. But also the inverse flow from production to innovation plays a crucial role, carrying over feedback on the adopted innovations, also providing stimuli for starting new innovation projects.

3.3. Business ecosystems and virtual enterprises

Business ecosystem and VE are two central notions in BIVEE. The notion of a VE refers to an organizational and business model where different production organizations (e.g. enterprises and production units (PUs)) join together with a predefined objective (e.g. achieving a given production or business innovation) and share skills and competencies in order to attain a specific result (e.g. product, service or a target market share) [AAG 99], Once the business objectives and the production plans are defined, a VE creates a VPS, i.e., a complex networked business environment where the value creation activities take place. A VPS is organized in different layers (e.g. operational, resources, management layer, etc.), where different resources are allocated (e.g. humans, services and machines) and different activities take place respecting different constraints and factors (e.g. production capacity, budget constraints, laws, deadlines, etc.)

The most important characteristic involved in the selection of a PU to participate in a VE is its capability. This characteristic reflects not only the availability of the required technology and relevant skill and experience, but also the reliability in terms of its records in respecting commitments, the capacity of producing the required volumes, the flexibility in relation to program changes, and so on. All this information is maintained in the business ecosystem repository (part of PIKR) and is constantly updated.

The coordination and collaboration of SMEs, or PUs, in a single VE may raise specific issues, arising from the fact that the different SMEs may significantly diverge in terms of organization model, management style, decision-making methods, information transparency (but also "hidden agendas"). Furthermore, different PUs and, more specifically, different production phases may be supported by different information systems (process models, data types, etc.). Such different information systems typically manage information having different representations, organizations, standards. Such a this diversity may cause that the information accompanying the outputs (products/services) released in one production phase is not fully (at least in an easy way) compatible with the input expected by PUs operating in the successive phase. To cope with this problem, BIVEE adopted a semantic interoperability solution, based on a central repository (including a set of ontologies) for the semantic annotation of the various resources by using a unified reference vocabulary and knowledge-based.

Furthermore, innovation requires continuous decision-making, often is to be made in a short time. Since there is no unitary center of command and there is a shared responsibility among the partners of the VE, there is the need to organize the VE information to support distributed decision-making according to non conventional methods. This is achieved primarily by providing solid, reliable fact and knowledge base: a prerequisite to achieve informed decision-making. The idea to measure, monitor and control innovation activities by using traditional time-cost-quality performance indicators is in general not suitable and in some cases could even be counterproductive, risking to seriously jeopardise an innovation project.

A second issue is a clear separation between local and global decision-making. BIVEE provides the required functionality to overcome this dichotomy, allowing a loose integration (i.e., just to the extent needed by an effective cooperation), where local modeling methods and working styles remain independent, but at the same time offering the solution to provide a global view of such a fragmented picture. Effective global coordination with marked local autonomy presents also the advantage of lowering the cost for a new enterprise to join a VE.

A Business Ecosystem (BES) [PEL 04] is a "protected" space populated by enterprises willing to stay in touch in the perspective to get together to set up a VE. To be accepted in a BES an enterprise needs to satisfy a number of criteria and to respect some common rules, once included in the community. Criteria and rules are freely decided in a democratic way by the community itself. Entering in the BES can happen by invitation or application. In both cases, the enterprise must provide a substantial information, including competencies, skill, capabilities, previous experiences, that composes the enterprise profile. Such profiles are freely accessible by the other members of the BES, and automatically filtered by dedicated BIVEE services that support partner search during the composition of a new VE. The BES rules define also the exit mechanism for enterprises intending to leave, including "non-disclosure" obligations regarding the other partners information acquired while participating in the BES. Among the various advantages offered by the participation in a BES, there is also the collective procurement opportunity.

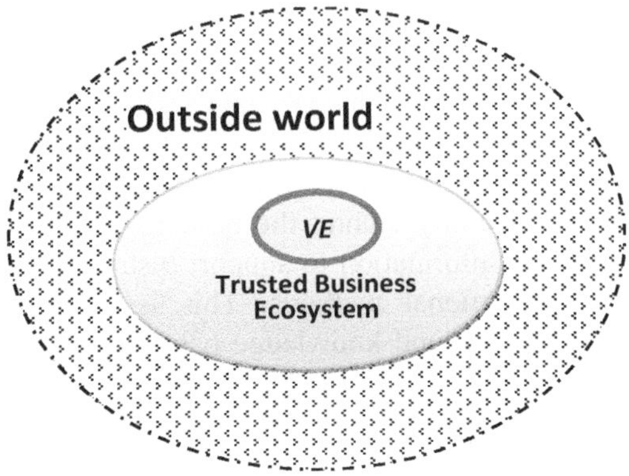

Figure 3.1. *A Participatory Space*

In a BES, enterprises and stakeholders operate in a participatory area (including both the production and innovation space) organized according to three concentric circles, as depicted in Figure 3.1. The *inner circle* includes the PUs that participate in the VE or VIF (several VE/VIF can be active in a BES at a given time). The *middle circle* represents the trusted business ecosystem (BES) that gathers all the enterprises that have been recognized and qualified, and are therefore able to guarantee the required levels of quality, performances, trust and security. The *outer circle* represents the "open sea", populated by all possible business players, potentially interesting for (and interested to) the ecosystem, but also player who compete in the same market and are potentially distrustful and hostile. The three circles offer different levels of openness and protection.

The key events of the BES lifecycle are listed below. BIVEE provides services to manage the following four basic events:

1) Evolution of the Business Ecosystem, when a new enterprise joins the BES or an existing one updates its capabilities or competences.

2) Exit of an enterprise from the BES, either for its explicit choice or for the progressive failing of the required membership criteria.

3) Start of a new VE, in response to a new market opportunity.

4) Termination of a VE having reached its business objectives, or for the failing to fulfill the conditions required by the BES charter.

As anticipated, the BIVEE philosophy is based on two interleaved spaces where value production and business innovation take place. In the next two sections we will introduce the two mentioned spaces.

3.4. Value production space

The advent of the Internet has already offered new opportunities for enterprises to improve their business and production models, one of the key impacts is represented by the increasing flexibility and adaptability of production carried out by a networked enterprise. Distributed, networked production models are difficult to design, implement, deploy and manage in an optimal way. The "traditional" approaches to enterprise optimization are not suited and new solutions are sought to manage networked production paradigms, capable of adopting the power of the Internet.

In BIVEE, we introduced the notion of a VPS, as a digital virtual reality aimed at modeling and representing a complex, distributed reality where a VE operates. In a VPS, we have PUs (PUs, sketchily represented by rounded boxes in Figure 3.2) that can be independent SMEs or organizational units in a single enterprise (we do not need to distinguish at this level). A PU can be of four different sorts: manufacturing, assembly, service and logistics. A production map is represented by a graph, with nodes that represent PUs connected by flow arcs. In a production map, that includes also other complementary infrastructures such as storage warehouses and other necessary services, there are branches with alternative or parallel paths (a path is a linear subgraph). The primary role of a value production map is not only to represent the flow of goods and services, but also other flows, such as the financial and information flows. A value production process evolves over the production map, traversing a number of (pre-) defined paths, links and units.

In our view, a VPS (and the VE operating therein) is not a closed territory: it is open for (impromptu) contributions from the rest of the BES and, in different forms, from the external world; at the same time it is sufficiently protected to ensure the required levels of trust and security to the members operating within such a space (as exemplified by the dotted border line of the production space in Figure 3.2 that represents a VE with its PUs). BIVEE provides the services and the methods to model the VPS, the VE, and to monitor the activities that take place along the production paths.

The MCR implements the services that BIVEE offers for managing the value production processes taking place in the VPS. In particular, it enables the users to:

– create and maintain the Business Ecosystem, that is the "incubator" of new VEs;

– support the creation of a VE, helping the discovery and selection of partners and possible evolutions of its composition;

– define and maintain the production map of each VE;

– monitor the distributed production activities, by gathering feed-back from each PU in the VE

– periodically update the production plan distributing the updated tasks and workload to the interested PUs in the VE.

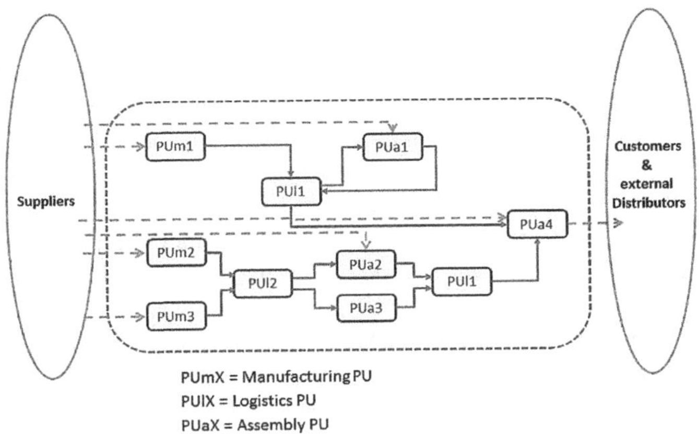

Figure 3.2. *A value production map*

As seen, while guarantees that the daily routine is constantly controlled for maximum adherence to programs, has the objective of improving current operations. This objective is supported by the production monitoring based on a number of *ad hoc* KPIs, tailored to each specific VE.

In setting up a new VE, BIVEE supports the development of two main documents:

– *Business Planning*, that is focused on the creation of the VE, starting from the production objectives, the market to be served and the production resources involved; it indicates also the expected revenues and costs. One of the main supports provided is the screening of the capabilities of the candidate PUs, in term of type of production, skills, acceptable volumes, time responsiveness, in order to configure the VE.

– *Master Production Planning.* This document comes afterword and concerns a given delivery plan; it is compiled after having verified the production objectives, with quantities and times, then the feasibility, by the configured VE, in terms of available resources, cost, schedule and so on.

Once the VE has been set up, the platform provides support to the continuous production monitoring and management activities. These have been formalized through a set of process templates (see Chapter 2) focusing on the production process as seen at VE level, where each component PU is seen as a work center of a traditional Enterprise Resource Planning (ERP) system. The support to the local production management and control system of each single PU is not considered part of the BIVEE mission.

The VPS processes are described according to the SCOR Framework[2] that introduces precise definitions of the functionalities mentioned above. To summarize, the primary production processes of the VE are (according to SCOR): *Value Chain Plan, Production Plan, Production Source, Production Build and Production Deliver*. These processes are applied to the VE in its entirety, but each PU will take care of specific tasks and will organize its own internal processes accordingly. The VPS processes, seen at VE level, are represented by networks of macro activities performed by the involved PUs. For example *Production Assembly*, an activity that can be found in the *Production Build* process, does not detail the specific sequence of operations in the local PU, but sees it as a "black box", considering in input the consumption of the required components, and in output the staging of complete assemblies, according to the Master Production Plan.

Plans that are distributed to the involved PUs can require some later negotiation in case of unpredicted events. For this reason, advancement feedback is harvested from the PUs, in the form of KPI measures, to create a picture of the progress in the production processes. In case that a PU is not able to send the required information, for instance, if its enterprise system is down, the BIVEE platform provides a Web-based user service for manual data collection.

---

2 Supply-Chain Operations Reference (SCOR) model: http://www.supply-chain.org/.

Production knowledge support

Focusing on the value production activities of a VE, a central role is played by the PIKR. The objective of this tool is to maintain a constantly evolving knowledge base, aligned with the state of play in the production field, extended along various dimensions:

1) Knowledge sources: the PIKR contains crowd-sourcing mechanisms, in order to harvest knowledge by an extended audience of business, market, technology experts, which may significantly help in production improvement.

2) Organizational memory: the history of all improvements and innovation exercises, whether successful or not, is maintained and can be retrieved in order to evaluate results and lesson learned.

3) Similarity of topics in different production activities: production improvement often arises from addressing the issues of a product or a process with solutions used in different domains. The availability of "knowledge nuggets" coming from more or less contiguous territories can trigger interesting searches of new materials, operations, applications, etc.

Business Ecosystem

To be considered as candidate for participating in a VE, an enterprise has to be screened on the base of a number of characteristics (reliability, technology level, production capacity, etc.) Enterprise profiles, maintained in the PIKR and adjusted to the evolution of the state-of-play, are analyzed in the context of the VE interests.

New members of the BES considered for membership after they explicitly apply (push mode) or by invitation (pull mode). The scouting of possible new entries in the BES is supported by the observatory function; when a new VE needs to be created and none of existing BES members is eligible, the VE manager can activate a BIVEE search to screen enterprises existing in the "outer world" that exhibit interesting characteristics. Out of the list provided by BIVEE semantic crawler, the VE manager can select candidates that can be contacted and invited to join the ecosystem first, and the VE afterwards.

Product and/or process improvement

VE changes aimed at achieving the planned improvements need to be implemented in a way that minimize the impact on BAU. Therefore, it is important to maintain a clear separation between current production programs and executions on the one hand, and feasibility of changes, assessing their effectiveness on the other hand.

The triggering of improvement requests, management of the changes, analysis of results, etc. are supported by the functions of the MCR, while the actual test of the implemented improvement, and the relevant feedback, is managed by the VPS monitoring services: the main knowledge objects involved in these services are KPIs, elaborated below.

Ongoing VE production

The SCOR modeling frame is the base of the MCR services that are grouped into the following components: *Modeler* aimed at the VE Setup, including the modeling of production maps in the VPS; *Assistant*, that supports the rollout of the Master Production Plan, with its production processes; *Monitoring*, that manages the feedback gathering from the operational field, based on the KPIs coming from the PUs. The knowledge objects involved in this case are on one side the production schedules for the PUs, and on the other the production advancement, reflected by the KPI measures, harvested from the PUs.

Figure 3.3 shows the relation of MCR to PIKR and the data exchanged with the PUs. Please note that such a data exchange actually takes place by using the RDM&S.

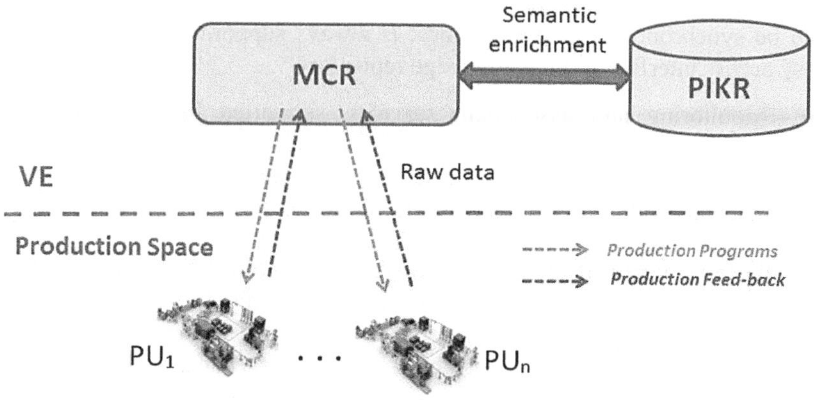

Figure 3.3. *MCR information exchange*

3.5. A participatory space for business innovation

One of the primary goals of BIVEE has been to address innovation with an "industrial" philosophy. In this perspective, innovation is seen as an intangible "knowledge artifact" to be built starting from "raw knowledge", suitably selecting and connecting the required "knowledge parts", then proceeding to create the "missing parts" needed to realize the final artifact, i.e., the body of knowledge that represents the sought innovation. The conception of such an engineering approach to innovation has been particularly challenging, since we had a narrow path to go, trying to push the proposed methodology toward a more rigorous and systematic approach, without jeopardizing the creativity required by innovation. Furthermore, the innovation process is inherently less defined than a production process. In this frame, we introduced the notion of a VIF, based on the following elements:

– Human experts, who are considered, with their intuitions and creativity, the primary engine of innovation.

– A set of guidelines, intended to be used by the innovation teams in a flexible and discretional fashion.

– A knowledge repository, holding all (possible) information that can be useful during the innovation project; it holds pre-existing knowledge, as well as the new knowledge produced (and acquired) during the work.

– A virtual collaboration space, where the distributed innovation teams can exchange ideas, questions, comments, answers. The collaboration, that can be synchronous or asynchronous, is always supported by a ubiquitous, easy access interface to the knowledge repository.

– Monitoring and assessment services, supported by a platform for collecting facts and evidences from the various innovation teams, feeding then the KPI Management System that will constantly provide the state of play of the project, based on quantitative elements.

The VIF eventually produces innovation artifacts that are passed to the VE and then applied to the VPS to concretely implement the innovative solutions. In rolling out the innovative solutions, the VE needs to carefully understand the cascading effects of the proposed changes. Such changes generally require the re-alignment of the VE over several dimensions and enterprise areas, beyond the one initially targeted by the sought innovation (here Change Management is centrally involved, not directly addressed by BIVEE since it is a well elaborated discipline, very rich in proposals and scientific results [PAT 08]).

In summary, business innovation is a designed, managed transformation of some aspects of an enterprise aimed at a substantial improvement of:

– the quality of delivery products (goods, services) and the customer satisfaction,

– production processes and workers satisfaction,

– cost reduction and/or revenue increase,

– sustainability of the production (i.e., implementing transformations respectful of working conditions, social context, environment, etc.) .

An innovation project typically starts with a (more or less) defined objective that is positioned in one focus dimension (e.g. introducing a new product in the catalogue), then it is necessary to identify all the other dimensions that are impacted by the specific innovation. Often, such an impact is overlooked and that implies a high probability of failure. The primary dimension that always is involved concerns enterprise processes, but to a certain extent all the other dimensions are affected.

The BIS is organized in a different way with respect to a VPS. The latter typically transforms raw material into finished products (or elementary services into complex services); while in the innovation space, we take existing production processes aiming at producing new processes. So, innovation primarily operates on the enterprise itself, that is the object of the transformation. Furthermore, an innovation process is inherently different from a production process, since the former is essentially ill-defined, typically built *ad hoc*. Even if it is based on existing guidelines and makes use of pre-existing practices (i.e., guidelines and subprocesses), a large space is left (hopefully) to creativity, intuition, “lateral thinking” that is hardly encapsulated in systematic, repetitive tasks. For this reason, we prefer to avoid the term “process” (implying a well defined and organized set of activities repeated in time) when we talk about innovation, “project” being more suitable.

An innovation project sees the initial involvement of creative units, followed by specialized units addressing specific issues, such as design, engineering, financial and market ones. All the units cooperate and interact by means of (generally distributed) communication platforms supporting the continuous exchange of ideas, comments and suggestions, design artifacts, mock-ups, blueprints, but also quantitative data and simulation outputs.

Similarly to what was introduced for a VPS, a BIS is characterized by an *innovation map*: a connected graph with nodes representing activities (and the teams enacting them) and arcs representing the flow of knowledge. As opposed to a production map, an innovation map is not fully defined when the activities begin, it is dynamically specified while it is traversed along the time, and new steps are dynamically identified, on the basis of the achieved results and the encountered problems. The traversed paths are aimed at progressively increasing the knowledge necessary to achieve the sought innovation. Such a progression typically starts from the existing state of play, a consolidated set of products and production process, gradually acquiring new knowledge until the innovative solutions are fully specified. In the most radical cases, the innovation project may require the extinction of (part of) what exists, to be replaced by radically new solutions[3]. In parallel, we need to consider the risk that an innovation project may fail, therefore it is necessary to show, with reliable forecasting methods, that the undertaken direction has good chances to terminate delivering the expected results.

Innovation management

At the core of the BIS is the concept of innovation management. Innovation management is the process of managing ideas through the stages of the innovation cycle until they reach an industrial maturity and eventually the market [HAM 07]. The innovation cycle describes the activities involved in taking an innovative idea, elaborating it to progressively achieve industrial strength products or services [OSU 08].

Generally the innovation cycle has been represented as a process flow taking place within a single enterprise, where the interrelations with the external environment are mostly marginal and external collaborations mainly provide punctual contributions. Such a form of "closed innovation" has been primarily motivated by the need to protect ideas and industrial innovation plans.

Starting from 2003, the paradigm of closed innovation has been debated and contrasted with the emerging paradigm of Open Innovation, first introduced by Chesbrough [CHE 03]. Open Innovation has nowadays taken a permanent place in the business and organizational world. Moreover, the advent of new ICT solutions and the predominance of Internet-based services

3 According to Joseph Schumpeter's Creative Destruction [SCH 94]

have made the enterprise environment progressively more open and participatory. One of the main issues of open, collaborative innovation concerns the protection of ideas and the intellectual property rights (IPR): there is the need of finding a virtuous balance between the need to disclose ideas and knowledge, to facilitate (even unexpected) external contributions, and to protect the key knowledge assets, to avoid that even partial results can be acquired and used by competitors.

In BIVEE, we proposed the business innovation reference framework (BIRF) to organize the BIS according a rigorous method while preserving the necessary flexibility. The BIRF is based on three main pillars: (1) the BIVEE Innovation Waves, (2) the document-based collaborative knowledge management, and (3) the monitoring strategy, based on carefully selected KPIs. All the three pillars are implemented by the collaboration space of the VIF (see below).

3.6. BIVEE innovation waves

An innovation project, typically triggered by the five anticipated cases (see section 3.2), starts by carefully considering new technological and market opportunities, having also in mind specific client needs. This is achieved with an exploration and scouting phase, including the access to pertinent knowledge available at well renowned research and innovation centers (of partners, agencies, universities, etc.) Then, according to BIVEE, an innovation project is organized following the four BIRF innovation waves: creativity, feasibility, prototyping and engineering, as described below.

Creativity Wave: This first wave starts with an innovation idea or a problem to be solved, providing a first description. Then, a team is established that starts to elaborate the initial idea, searching for similar ideas and previous experiences (including past failures) related to it. If the idea is promising, a VIF is created, with one or more teams with resources belonging to different real enterprises.

Feasibility Wave: This wave starts when the initial idea is sufficiently specified, with its structure, functions, scope and intended impact, and approved by the corresponding manager. Then it proceeds elaborating the business plan, market analysis (including competitors, customer profiling) and the feasibility documents (e.g. technical, financial, industrial, market feasibility), with risk analysis and patent possibilities. The output of this wave, that includes a preliminary design, is checked by the top management that, in case of positive evaluation, allocates the required budget.

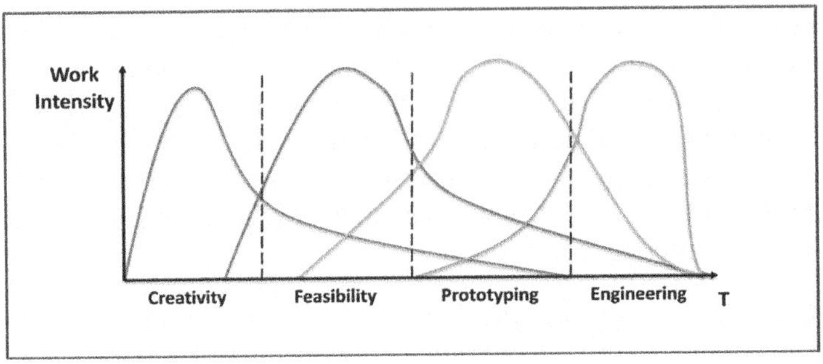

Figure 3.4. *The four innovation waves of BIVEE*

Prototyping Wave: This wave starts elaborating the detailed design and proceeds with the first implementation of the initial idea. Simulations and extensive testing (in the lab, as well as in the field) are carried out. This is the wave in which the innovation is concretely projected into the real-world for the first time, allowing the initial idea to be confronted with implementation and practical issues.

Engineering Wave: This is the final wave that concerns the industrialization of the innovative solutions, with the documentation to be transmitted to the VE to start the rollout of the innovative solution in the VPS. Also the prospective budget, break-even point and business model are defined, together with all the other documents that concur to form the knowledge asset necessary to activate the industrial production. Such knowledge includes documents like the Bill of Material (in case of a manufacturing product), production plans (indicating what to make or buy), delivery strategy and set-up instructions, testing and maintenance procedures.

We call them waves since they are logically sequenced in the time, but they are tightly interconnected and the start of a new wave does not imply that the previous wave has been fully accomplished. Furthermore, there will often be the need to jump back and forth to complete a document or to correct it on the bases of later findings. For instance, during the prototyping wave there can be new findings that jeopardize the results of a previous financial feasibility study, requiring therefore to rethink some parts of the innovation under elaboration. The wave approach is depicted in Figure 3.4.

The innovation waves represent a powerful framework to guide innovators in achieving an innovation project. They are sufficiently powerful to be used for different kinds of innovation (e.g. product, process, market, etc.) and in different application domains (from automotive to tourism, from health to Government.) The guidelines associated with the waves indicate a number of document templates that, to be filled, require specific knowledge to be gathered. Waves are also associated with a number of carefully conceived KPIs and a method to monitor and assess the progress of the work.

The architecture of the VIF platform

The BIRF framework represents the rationale of the VIF, and in particular of the software platform that supports the innovation projects. The VIF platform has been conceived to support collective knowledge creation and management of innovation. To this end, the primary components are:

– *Shared Semantic Whiteboard* (SSW): This is a knowledge sharing and visualization Web platform, where each member of the innovation team can post ideas, comments, issues, etc. The SSW can be remotely accessed, it displays elements that can be seen, edited, commented by all the members of the team (both in sync and async mode). Furthermore, the platform proactively searches and associates to SSW elements relevant knowledge items (e.g. documents) extracted from the Open Innovation Observatory and the PIKR (see below).

– *Open Innovation Observatory* (OIO): This is a knowledge repository where the dedicated search engine of the VIF collects and organizes the material extracted from various public sources, both on the web (the open section) and on the partners local repositories. Among the stored knowledge, there is information about worldwide excellence centers, various projects and scientific results, relevant for the community. A careful and continuous observation of the evolution of the external world, managed by OIO, guarantees that BIVEE holds the information necessary to timely trigger suitable optimization and/or innovation actions.

– *Collaborative Innovation Capability Maturity* (CICM): The BIVEE framework has also proposed a set of criteria and guidelines for networked SMEs (and enterprises in general) to (1) verify the innovation readiness of the enterprises and (2) to identify a route that an enterprise can follow to progressively improve its innovation capability. The innovation CMM Model is inspired by the well known capability maturity model originally introduced for the Software Engineering by the Software Engineering Institute (SEI), but in addition to the five levels that concern the CMM of a single enterprise (Initial, Repeatable, Defined, Managed, Optimized) it introduces another dimension to cater for the virtual (networked) aspect of the VE. The latter dimension is in turn organized in four levels: single organization, network awareness, network consent, network dedication. The CICM model has guided the implementation of a web-based tool for the self assessment of innovation maturity of a VE (http://innonetscore.de).

3.7. An integrated view of VPS and BIS

Business innovation and production improvement both have in common the fact that they aim at changing for better some aspects of the enterprise. As anticipated, a crucial problem is the identification of the key enterprise dimensions where such changes will take place. It is obvious that the enterprise dimensions are not independent on one another and, in general, the first changes will trigger a sort of chain effects, with a more or less wide impact. In a broad view, the enterprise transformations can involve the following dimensions: *Product, Service, Process, Technology, Organization, Market and Strategy.* In BIVEE, we decided to focus on the first four dimensions.

Today, the transformations are increasingly happening at a fast pace, with a trend toward a continuous improvement/innovation paradigm. Figure 3.5 shows a representation of two nested cycles, one concerning the optimization cycle and the other the innovation cycle.

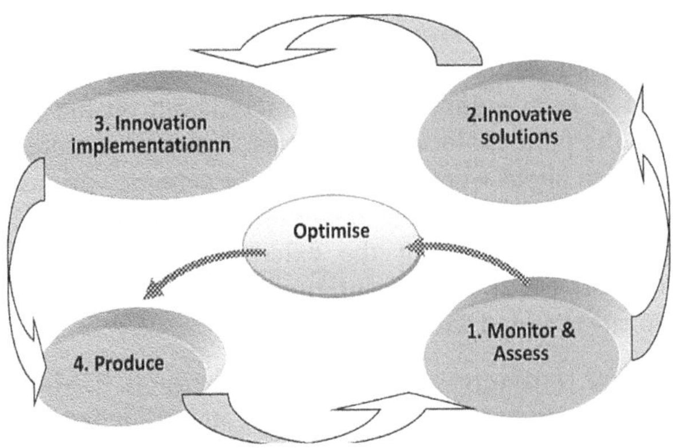

Figure 3.5. *The optimization–innovation cycles*

According to the BIVEE approach, the optimization–innovation cycles are tightly interwoven, supported by the two application platforms that we have seen above: the VIF and the MCR. In fact, the two corresponding spaces, BIS and VPS, are also tightly interwoven, while maintaining respective roles, objectives, and characteristics that are inherently different. As anticipated, the idea is that an innovation map takes as input a "consolidated" value production map and generates as a result a new, innovative, production map.

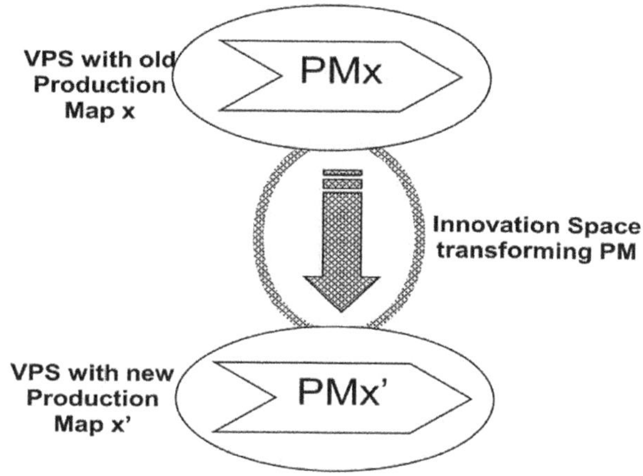

Figure 3.6. *Integrated production–innovation space*

Figure 3.6 represents such an integrated view, where the upper part shows the existing production map (PMx), i.e., the AS-IS situation that is going to be changed by the sought innovation; the central part symbolise the innovation space where an innovation solutions is progressively elaborated; the lower part illustrates the innovated production map (PMx'), after the innovation has been fully implemented. The BIVEE platform essentially operates on this metaspace, adopting the most effective knowledge representation methods and notations.

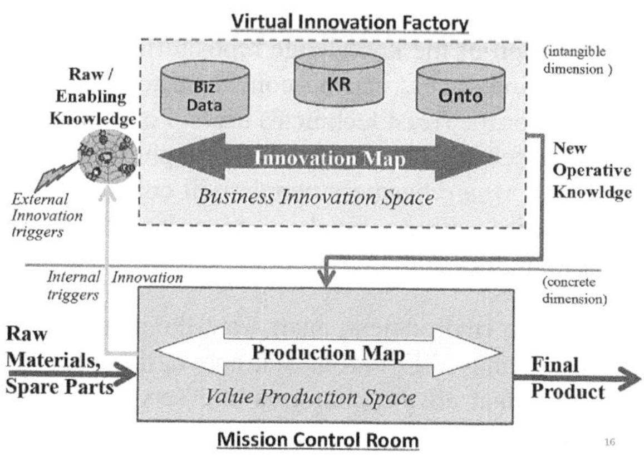

Figure 3.7. *The interplay between VIF and MCR*

3.8. The macro-architecture of the BIVEE platform

In the foreseeable future, there will be a strong impact on organizational models where enterprise application and services will undergo a transformation of the traditional monolithic on-premise systems (such as ERPs) into heterogeneous collections of services, available from different providers on-demand and, e.g. on a pay-as-you-go basis. The traditional three-layer architecture, *data–application–presentation*, is now implemented by services which allow for a boundless access to heterogeneous data (e.g. with a linked open data approach), for the accomplishment of enterprise operations (planning, forecasting, monitoring and management) and the implementation of new forms of mobile and fixed human–computer interaction (the so-called *Service Front Ends*). *Cloud computing* is also pushing forward a comprehensive service paradigm to implement infrastructure-, platforms-, and software-as-a-Service (IaaS, PaaS and SaaS) solutions. Finally, some functions previously implemented in the application part of the enterprise system are now moving to the infrastructural part (see, for instance, the idea of an ISU: *Interoperability service utility*), becoming new kinds of network utilities. We envisage that business intelligence, and more generally Big Data Analytics, services will be among the key enablers for a virtuous growth of EU enterprises and will be among the value-added services that will characterize next generation enterprise applications capable of influencing new styles of fact-driven management.

In this frame, the macro-architecture of the BIVEE Platform is conceived to put the business user in the center. In particular, the idea is that managers will progressively become "coaches", active in monitoring and managing both production and innovation spaces with an open, participatory style. To this end, the organization of the knowledge repositories, the offered services and, overall, the user interfaces, will be conceived to create a comfortable, familiar environment for the (least technical) users. For this reason, particular attention has been placed to achieve advanced graphical user interfaces for the MCR and the VIF, where business people will collaborate in monitoring and managing enterprise entities (much in accordance with the principles indicated FInES Research Roadmap 2025[4]).

The BIVEE software environment, apart from the platforms conceived for the end users, MCR and VIF introduced above, includes the following platforms. Please note that all the platforms and services introduced in this

4 Future Internet Enterprise Systems Research Roadmap. EC Task Force Report. http://cordis.europa.eu/fp7/ict/enet/documents/fines-research-roadmap-v30_en.pdf.

chapter will be addressed in the book by the chapters specifically dedicated to each of them.

Production and Innovation Knowledge Repository (PIKR): This repository contains the data and knowledge concerning production and innovation activities coming from the different partners of the VE and VIF, with the objective to provide a wide but “controlled” knowledge sharing. It complements the OIO that, conversely, holds public “external” knowledge. In particular, PIKR stores knowledge about all the teams working in the BES with their competences and experiences, the achievements of past projects, but also information of the VE production space evolution. PIKR is centrally based on the various ontologies: *DocOnto*, for the business document templates, *ProcOnto*, for business process schemes, and *KPIOnto*, for the various KPIs. The PIKR, that adopts particular security mechanisms to protect its content, is a virtual repository since the actual resources physically reside locally, at VE/VIF’s partner’s sites: it hosts and manages ontology-based images of such resources.

Production & Innovation Key Performance Indicators (PI-KPI): One key role of BIVEE is the monitoring of the achievements of the PUs working in a distributed, collaborative fashion. To this end, the project proposes various KPIs, both for production (P-KPI) and for innovation (I-KPI). There is a rich literature proposing a large number of performance indicators, different for goals, domain of interest, degree of precision, and formalism, coming, e.g. from international and national standardization bodies, including reference models like the Supply-Chain Operation Reference (SCOR) model[5] and the Value Reference Model[6] (VRM). KPIs in BIVEE are evaluated starting from the information coming from the field operations and represent an invaluable asset on which informed decisions can be made. The relevant KPIs are organized in a hierarchical structure, as specific indicators can be grouped in a higher level ones: a specific tool, the KPI Modeler, enables business experts to easily create these structures. The monitoring framework (supported by the *KPIOnto* ontology) currently holds a total of 69 KPIs for the production space, 100 KPIs for the innovation space and 29 KPIs concerning the sustainability context in both spaces.

Raw Data Management and Service (RDM&S) platform. This is the enabling platform that connects the various information sources, local to the different enterprises cooperating in the VPS or the BIS.

5 http://supply-chain.org/f/SCOR-Overview-Web.pdf.

6 www.value-chain.org/value-reference-model.

The RDM&S platform guarantees a constant surveillance of the ongoing activities, within the MCR and VIF, collecting and homogenizing the key performance measures (KPMs) that represent the factual evidence necessary for deriving the KPIs.

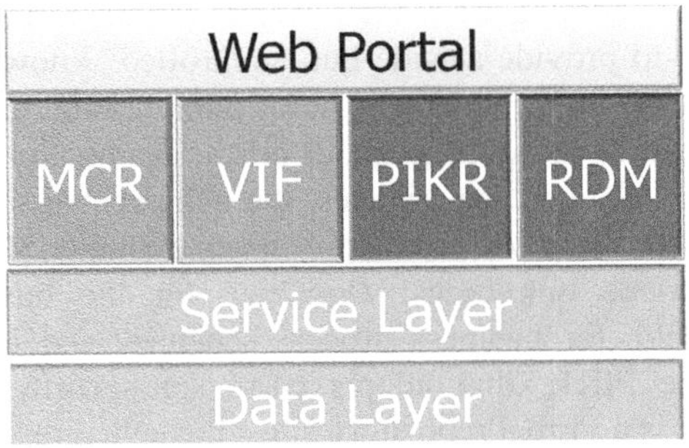

Figure 3.8. *The BIVEE runtime framework*

Figure 3.8 illustrates the global architecture of the BIVEE software environment. It highlights the web-based user interface on the top and the service-data layer on the bottom. In the middle, the four software platforms that provide the end-user services (the two blocks on the left) and the knowledge-data services (the two blocks on the right.)

3.9. Trial cases and impact

Impact creation has been one of the central goals of BIVEE and therefore an integral part of the project. To this end, two pilots have been carefully selected to provide a good coverage of different industrial cases; in fact, they are positioned at the two extremes of the technological scale: one belonging to the low-tech sector of wood and furniture and the other to the hi-tech sector of robotics and automated measuring equipments. This strategic choice aimed at demonstrating that the value proposition of BIVEE can really have a wide scope. The second qualifying choice has been to carry out a comparative assessment in two distinct phases, launching two different monitoring campaigns. The FMC aimed at acquiring evidence about the performance of production and innovation activities in the "as is" situations before the adoption of BIVEE. The FMC, apart from collecting quantitative and qualitative information from the operational fields, brought to the project the strategic advantage to involve end-users as early as possible. Getting in contact with the future adopters of BIVEE gives us the possibility to introduce them to a number of clues, while allowing us to gain a better understanding of a number of issues that were scarcely considered in the initial design of the software environment. Such issues, if underestimated in the early design phase, would have caused a significant delay in the later phases of the project. This has been an important lesson for the engineers and designers, understanding that a good expertise and a strong theoretical background need to be always confronted with the real needs of the users and the stakeholders.

After the FMC, we finalized the first BIVEE prototype, starting to progressively deploy it, in order to begin the SMC aimed at gathering the information on the two trial cases observed in presence of the BIVEE platform. The objective has been that of confronting the evidences collected in the SMC with the corresponding evidences collected in the FMC. As usual, the reality fully reveals its complexity only when you start to practice it, any kind of theoretical model will be hardly able to seize it. Therefore, in the SMC, we realized that a systematic contrast with the FMC data could be hardly achieved. The reality proved to be faster and slower than expected. Faster, since in a year many things have changed even in the same industrial context (due to the international crisis, to split and merge of enterprises, etc.) and slower, since we realized that many enterprise phenomena, especially those connected to the structural changes, need time to take place. Therefore, a monitoring campaign of six month is too short to cover a significant time span, both in case of improved production processes and innovation projects. Nevertheless, we were able to collect several evidences showing the positive impact induced by the adoption of BIVEE. At a strategic level, the large majority of users and stakeholders agreed that BIVEE is going to occupy an important area where there is very little offering and, at the same time, where there is a manifest need. Such a need will be steadily growing in the future, since even when the economy (hopefully) will restart, the call for more effective management of production improvement and innovation will increase. And solutions like BIVEE will be needed furthermore.

The main lesson learned concerned the limit of technology, especially if introduced in a working context where there is still a digital gap. There is a risk embedded in the idea that good technical solutions will "automatically" improve to way people operate, make decisions, collaborate, produce value. We already knew it, but we had a further evidence that when we deal with sociotechnical systems, such as the BIVEE Environment, the "socio" component plays a central role. Therefore, when designing such a kind of systems, in parallel to the development of the technological solutions, it is necessary to consider, with a marked collaborative and participatory approach, also the new socio-organizational solutions, including educational initiatives for the involved people.

3.10. Bibliography

[AAG 99] AAGEDAL J.O., MILOSEVIC Z., WOOD A., "Modelling virtual enterprises and the character of their interactions", *Research Issues on Data Engineering: Information Technology for Virtual Enterprises*, RIDE-VE '99, 1999.

[KNO 12] KNOKE B., State of the Art about the Collection and discussion of requirements for Systematic Business Innovation and definition of KPIs, BIVEE Deliverable D2.1, 2012.

[CHE 03] CHESBROUGH H., *Open Innovation: The New Imperative for Creating and Profiting from Technology*, Harvard Business School Press, Boston, Massachussetts, 2003.

[HAM 07] HAMEL G., "Management innovation", *Journal on Leadership Excellence*, vol. 24, No. 1, p. 5, 2007.

[MAR 01] MARTINEZ M.T., *et al.*, "Virtual enterprise – organisation, evolution and control", *International Journal of Production Economics*, vol. 74, nos. 1–3, 2001.

[MIS 12] MISSIKOFF M. (ed.), "Future Internet Enterprise Systems (FInES) Research Roadmap", Report of EC FInES Research Roadmap Task Force. Available at http://cordis.europa.eu/fp7/ict/enet/documents/task-forces/research-roadmap/fines-researchroadmap.pdf , 2012.

[OSU 08] O'SULLIVAN D., DOOLEY L., Applying Innovation; SAGE Publisher, 2008.

[PAT 08] PATON R.A., MCCALMAN J., *Change Management: A Guide to Effective Implementation*, SAGE 2008.

[PEL 04] PELTONIEMI M., VUORI E.. "Business ecosystem as the new approach to complex adaptive business environments", *Proceedings of eBusiness Research Forum*, 2004.

[SCH 94] SCHUMPTER J.A., *Capitalism, Socialism and Democracy*, Routledge, London, 1994.

PART 2

Storytelling on BIVEE Experience

4

A Project of Collaborative Networked Innovation

This chapter is in the from of a narrative, describing the possibilities offered by BIVEE in monitoring and contributing to innovation with ideas and expertise, in a networked context. This is an example of a typical innovation project, taking place in a realistic industrial scenario conducted through the BIVEE platform. The project deals with the development of a new sensor focused on a Fiber Bragg Gratings (FBG) system for real-time monitoring of the bridge pier for the scour risk during a flood. In particular, the innovation project starts with a specific request from Loccioni's top management and the idea proposed in order to measure the scour risk. Then the key players are introduced, followed by a synthetic description of the scenario. Finally, the innovation waves are illustrated and suitably instantiated for the specific innovation project[1].

4.1. Introduction

In an internal meeting, Loccioni's top management asked the innovation team to study a monitoring system that is able to measure and control the level of safety of structures, in particular that of bridges. The need is to be able to give warning and make decisions on the security levels of a structure like a bridge based on different types of measurements. Actually, the decisions are made based on the subjective experience of technicians and this

---

Chapter written by Cristina CRISTALLI, Daniela ISIDORI and Isabella TERZONI.

1 BIVEE Deliverable D7.40 – Comparative Impact Analysis of BIVEE Pilot Applications. http://wordpress.BIVEE.eu/resources/publications/.

approach needs to be more robust and substantiated by data and objective measures. In particular, it is important to study the effect of a dynamic system, such as the Esino river, on the Scisciano bridge (Maiolati Spontini, AN) situated along the street that connects two buildings of the Loccioni Group. In fact, the river is very close to the headquarters of the Loccioni Group and it is a precious energy source, but also a source of risk; therefore it has to be continuously monitored through a properly designed monitoring system.

One of the main causes of bridge failure is the local scour around bridge piers by river floods. There is experimental evidence that bridges collapsing during floods is not so unusual. An example is the Scisciano Bridge: in 1990, this bridge partially collapsed during a flood of the Esino River, precisely from pier scouring problems. Unfortunately, models for accurate and robust predictions of the scour levels are not available. Bridge builders overcome this deficit by using very large safety factors that are not effective.

The project has been named Flumen and it entails the development and set up of a monitoring system in order to control the phenomenon of pier scouring. The fiber Bragg gratings (FBG) monitoring system is a sedimeter consisting of an array of FBGs. The developed sensor is an improvement of a system patented by a University[2] and it is able, due to a new heating circuit, to detect the difference of environmental temperature among water and saturated soils, measuring the different thermal exchanges in order to determine the level of the river bed during a flood. If the level of the liquid–solid interface locally decreases due to the scouring, the system is able to detect this change and to give an alert in the case that the bridge is in danger.

Hereafter, the different waves managed through the BIVEE platform are illustrated in order to describe the benefits of the proposed approach and tools in the management of the Flumen innovation project.

4.2. Creativity wave

4.2.1. *Proposed idea*

In an internal meeting, Loccioni's top management asked the innovation team to study a monitoring system able to measure and control the safety

2 Due to a non-disclosure agreement signed, we can not specify the name of the University.

level of Scisciano's bridge, checking for seismic vulnerability. Daniela, Ph.D. research engineer for the innovation department, has an intuition: the vulnerability of Scisciano's bridge may not be related to seismic risk but to hydro-geological risk. She wants to share this idea with other engineers.

It is fundamental to be able to share the initial idea and all the related information with other members of the group, select the competent people as soon as possible and create a dedicated team able to develop the research project in an effective way: the innovation team. An "innovation team" is a dedicated group consisting of members, possibly coming from different enterprises, who interact and collaborate in order to speed up and carry out an innovation project.

Daniela logs into the BIVEE environment (Figure 4.1), inserts the idea in the virtual innovation factory[3] and starts her searches (Figure 4.2). First of all, once she has checked the notification of the dashboard, she inserts the issue, i.e. a short description of the idea and its requirements, together with the benefits both for customers and for virtual enterprises. Flumen is the title of the idea.

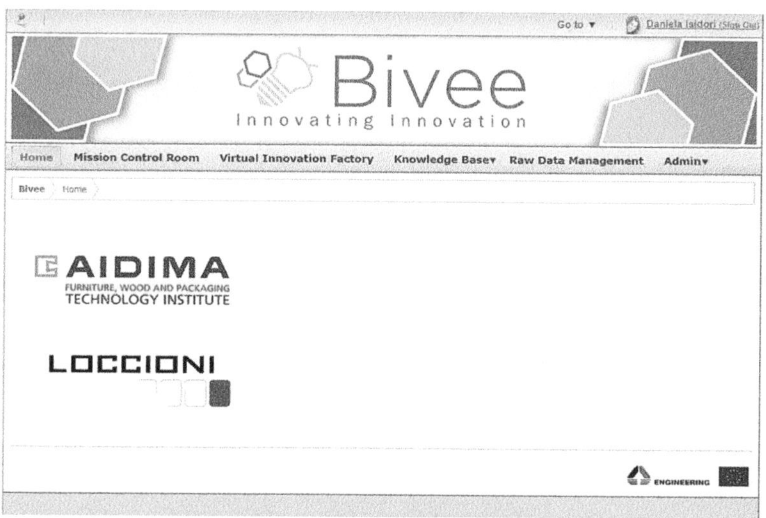

Figure 4.1. *BIVEE platform interface*

3 http://BIVEE-sp.eng.it/web/guest/home.

The proposed monitoring system of the hydrogeological risk of Scisciano bridge should be able to understand in real time with an automatic measuring system, the stability of the bridge and eventually to take prompt decisions on its eventual closure. There are several benefits. The benefits for the customer are related to giving support to the ordinary and extraordinary maintenance in case of emergency, increasing the security for landslide risk. The proposed system should be applicable to more than 30,000 bridges in Italy alone. With respect to virtual enterprise, the benefits are related to exploring a new application field, with the possibility to study new solution and to create new business.

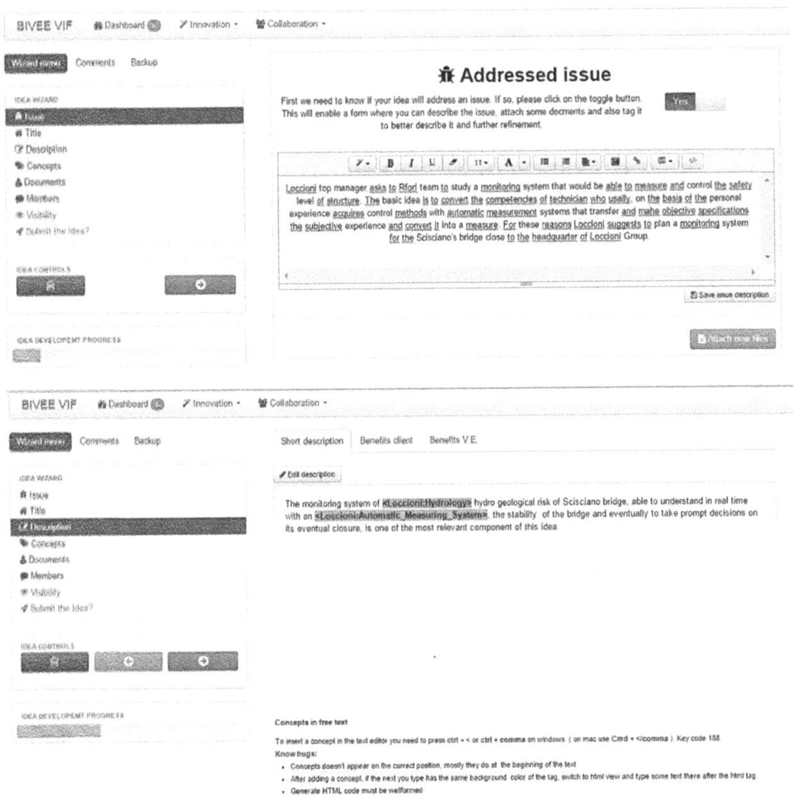

Figure 4.2. *Issue insertion and description*

During the insertion of the description, Daniela inserts some characterizing concepts, for example *<Loccioni:Hydrology>*, *<Loccioni:Automatic_Measuring_System>* annotating and highlighting them in the text (Figure 4.2).

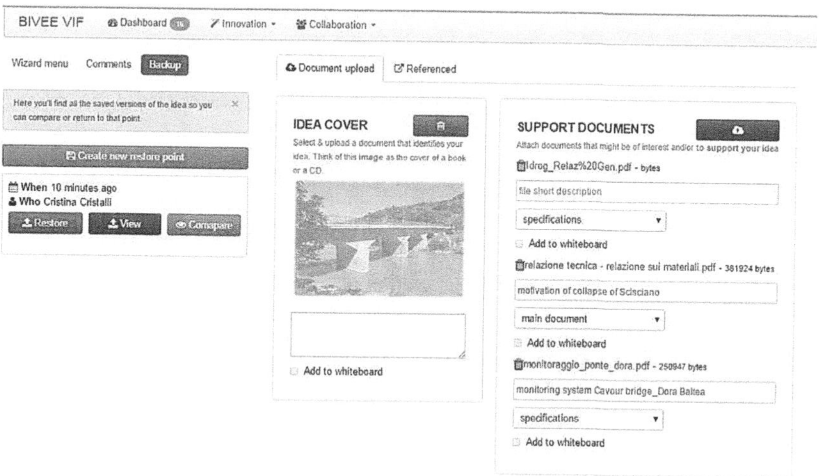

Figure 4.3. *Documents and idea cover in idea wizard*

Daniela already knows that the Scisciano bridge partially collapsed in the 1990s during a flood of the river Esino and she tries to discover the reasons why. The documents obtained from the web searches are uploaded in the document page of *idea wizard* (Figure 4.3).

She shares the collected documents and selects the *domain* (metrology), the *concepts* (automatic measuring system, measurement procedure, real-time monitoring system and sensor) and the *skills* (knowledge and/or attitudes) required to develop this idea (Figure 4.3).

By adding these competencies, it is simpler to select the researchers who will contribute to the development of the project or who have experience in the specified fields. In this way we can save time in the setting up of the working team and we can have benefits from existing experiences.

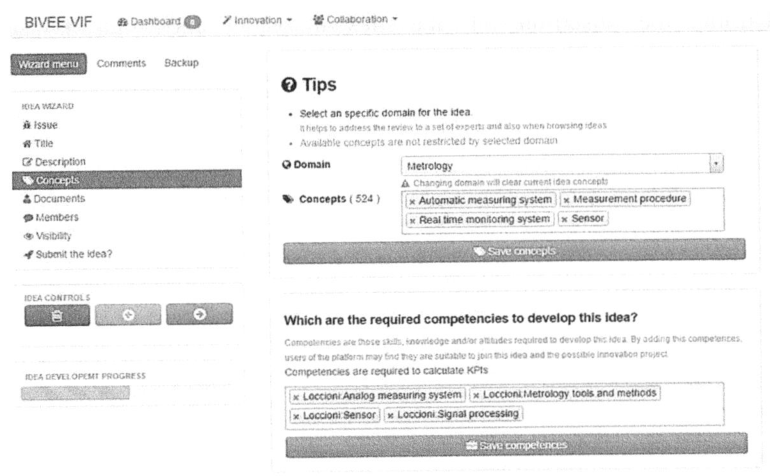

Figure 4.4. *Domain, concepts and competencies tool*

Using the search-by-skills, she selects three possible required competencies for this idea: metrology, signal processing and sensors. In this way she finds Enrico, a researcher in the research for innovation group who is a domain specialist of metrology and sensors; and Cristina, the head of research for the innovation group, with experience in signal processing (Figure 4.5).

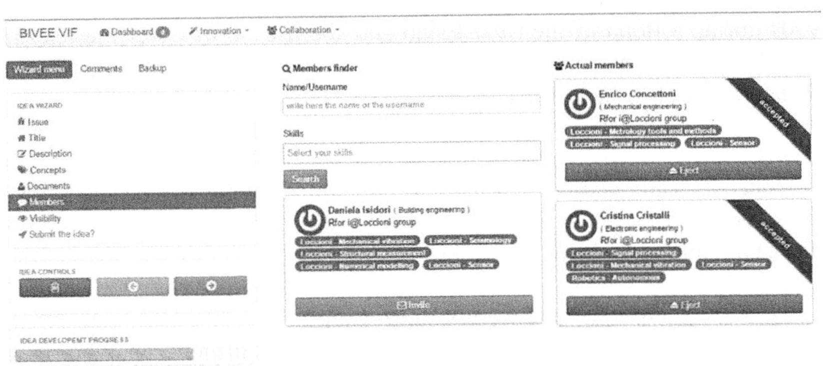

Figure 4.5. *Member's tool in idea wizard*

The VIF platform sends an invitation to each of the two BIVEE users through an e-mail automatically. Once the users accept the request, their status changes and they can visualize the idea.

At this point, Daniela selects the level of visibility that the idea should have once submitted.

The *visibility level*, that identifies the group of users that can visualize the idea after the submission, can be chosen in this wave coherently with the decided open innovation approach. The idea can be shared with the virtual enterprise, the business ecosystem or it can be public.

She decided to share the idea with all the members of the Loccioni virtual enterprise already selected in BIVEE platform. The Loccioni virtual enterprise consists of four Italian universities and two private enterprises.

Then, the new innovation team, consisting of Daniela, Enrico and Cristina starts to look for information about hydrogeological risk and measuring system. There are many existing solutions able to measure capacity flow, velocity and height. In the internal database of the Loccioni group, Daniela finds the *technical report* in which it is reported that the bridge collapsed due to pier scouring problems. One of the main causes of bridge failure is the local scour around bridge piers by river floods (Figure 4.6).

a) b) c)

Figure 4.6. *Pier scouring phenomena*

Unfortunately, models for accurate and robust predictions of the scour levels are not available.

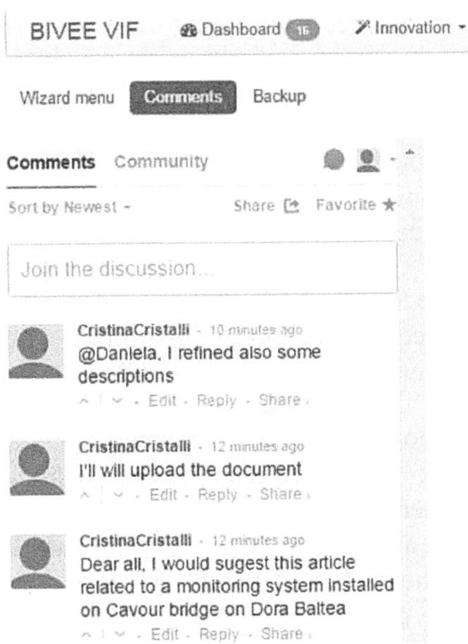

Figure 4.7. *Comments in idea wizard*

The innovation team uses the *idea wizard* (Figure 4.7) to upload documents and add comments (it is a sort of remote desktop where each member can modify, add, share documents and interact) and the whiteboard (Figure 4.8) in order to organize all comments and relevant documents related to pier scour problems due to flood.

These tools and methodology are very useful in order to avoid sending e-mails with big attachments and to have a common space for sharing useful documents. The sharing of comments within an innovation team is also very effective in terms of collaboration speed.

The web search results provided two different technologies that are used in order to measure the pier scour and they are based on sonar and FBG. The first technology consists of the installation of an echo sounder on the pier of the river in order to monitor the river bed variation. The second technology indeed is a promising technology based on an array of FBG temperature sensors capable, due to a heating circuit, of measuring the different thermal exchange between water and saturated soil and to determine the liquid–solid interface (river bed). Daniela also discovered that the sedimeter based on the FBG technology is covered by a patent owned by a university.

In order to always have a backup of all the information inserted into the idea wizard during the elaboration of the draft, all the modifications made can be saved creating a *restore point* that can be viewed and compared with the others. This is one of the fundamental aspects related to knowledge management in the BIVEE project: it is very important to observe and save all the information related to a project in order to obtain a complete repository.

Together with the *restore point* the VIF allows to evaluate the quality of the idea in term of quantity of information provided by the *idea quality meter tab*. Each member can check the idea quality meter: if the idea reaches the minimum quality for the submission, then the idea can be annotated, otherwise, the platform avoids the possibility to move forward.

In the present scenario, all the requested fields have been completed and many documents uploaded. Therefore the platform allows the next step: the *idea submission*.

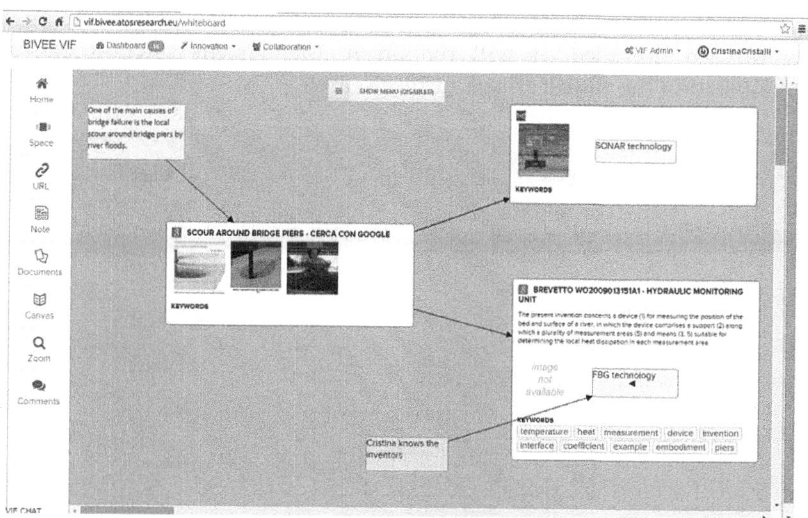

Figure 4.8. *Whiteboard*

4.3. Idea submission: Flumen

Once the idea reaches a good level of completeness, it is ready to be submitted and shared with the members chosen in the visibility tab and to be reviewed by all the users logged as "domain expert".

Daniela compiles a short description introducing the idea to domain experts who will receive a notification in the *dashboard* (Figure 4.10).

This aspect is fundamental because it allows her to find new enhancements and suggestions for the idea provided following an open innovation approach and to use the experience of a specific expert to speed up the evaluation of the idea. The domain experts in fact play an important role in the idea management; they are automatically advised with a notification of the submission and they can promptly evaluate the innovation potential of the idea (Figure 4.10). The idea has to be annotated before its submission (Figure 4.9). In order to accelerate the submission, some information areas are automatically extracted in order to fill some fields of the annotation template and to automatically create the *proposed idea* document (if this idea is accepted by the domain expert). Once the idea has been delivered to the platform, the domain experts will evaluate it and they eventually can ask for improvements. If they discard the idea, it will remain in the system since it can be re-evaluated in the future after a further search.

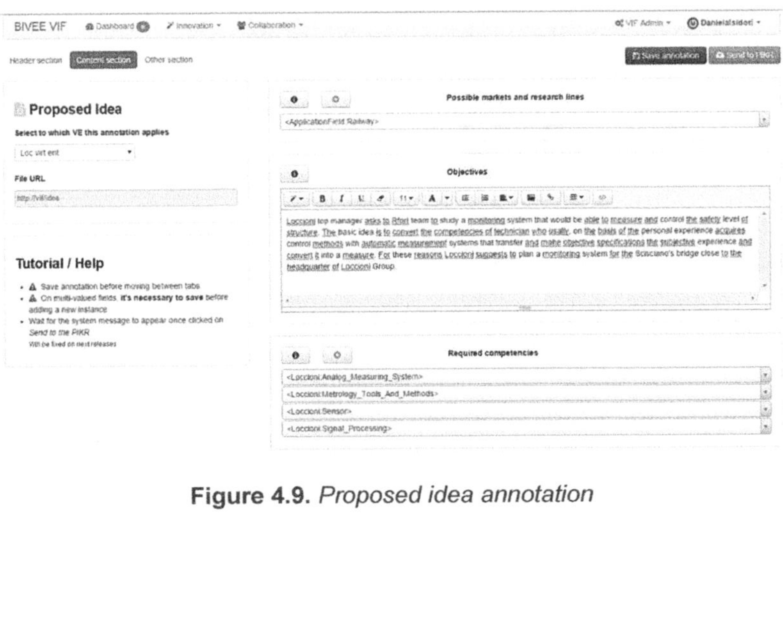

Figure 4.9. *Proposed idea annotation*

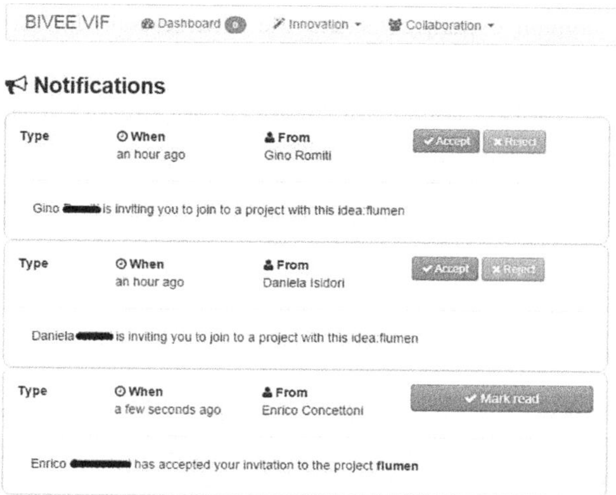

Figure 4.10. *Dashboard notification*

While the notification is sent to all domain experts of the BIVEE platform, the submitted idea is posted to the ideas wall (Figure 4.11). The *visibility* of the posted ideas depends on the *visibility* chosen in the *idea wizard*. So, in this case, all the members of the Loccioni virtual enterprise see this idea together with domain experts (in Figure 4.10, the *idea notification* for the domain expert). The *idea notification* is also sent to the VIF administrator (as you can see in Figure 4.10). This is another important role because he is responsible to check the operation conducted in the VIF. In any case, if the idea is rejected, the motivation for this rejection needs to be provided as feedback.

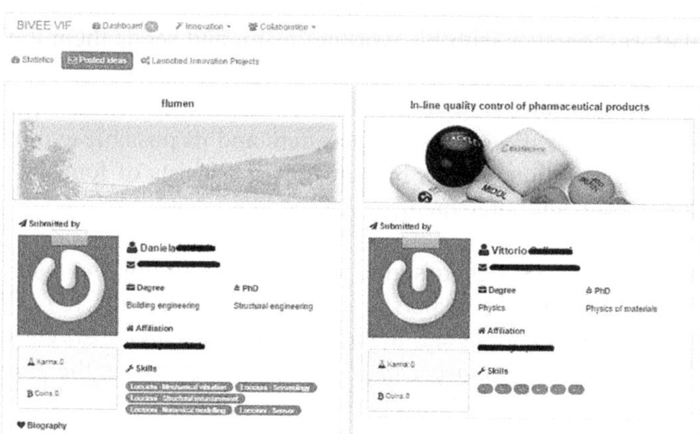

Figure 4.11. *Posted ideas*

Gino, as domain expert, considers the idea suitable for an innovation project and he gives his "ok" for launching the project.

Having launched the *innovation project*, Daniela selects the ontology elements for KPIs calculation (see Chapter 2 for references). As the first step, Daniela assigns a name to the project, which is used by other subsystems of the BIVEE platform to correctly refer KPI values. The chosen name of the current project is "*Flumen*".

4.4. Innovation project: Flumen

To start a new *innovation project*, it is required to have a posted idea. Once a positive evaluation of the idea has been obtained, the project starts. At this point, the innovation project related to the specific idea can be shared only with the selected members of the project. The VIF automatically sets up the *Flumen Innovation Project*, opening all the document templates that will be progressively filled with the acquired/generated knowledge. The *Flumen Innovation Project*, aiming at developing an improved version of FBG system, is then launched. It proceeds along the *4 BIVEE Waves*: *creativity* (already started), *feasibility*, *prototyping* and *engineering*, according to the BIVEE- Business Innovation Reference Framework (BIRF) (see Chapter 5 for references).

At this point, Cristina, as head of Research for Innovation Department, contacts the owners of the FBG system patent. The University is already part of the Loccioni virtual enterprise because it had already collaborated with Loccioni in the past. Cristina searches the name of the Professor in the BIVEE platform together with his working team, and sends them the request for the participation as members of the Flumen Innovation Project. A special agreement has been prepared in order to be able to use the patent for the possible commercialization of the FBG system and its possible evolutions. At this point, the innovation team is bigger: now it consists of four members of the Loccioni Group and three members of the University.

In the mean time, Daniela assigns, using the Gantt tool, the data for each task (Figure 4.12) and each person responsible (Figure 4.13). Enrico is responsible for the uploading and the annotation of *technical solution report*, Gino for the *preliminary market analysis*, and Cristina for the *prospective budget* and *innovation report*.

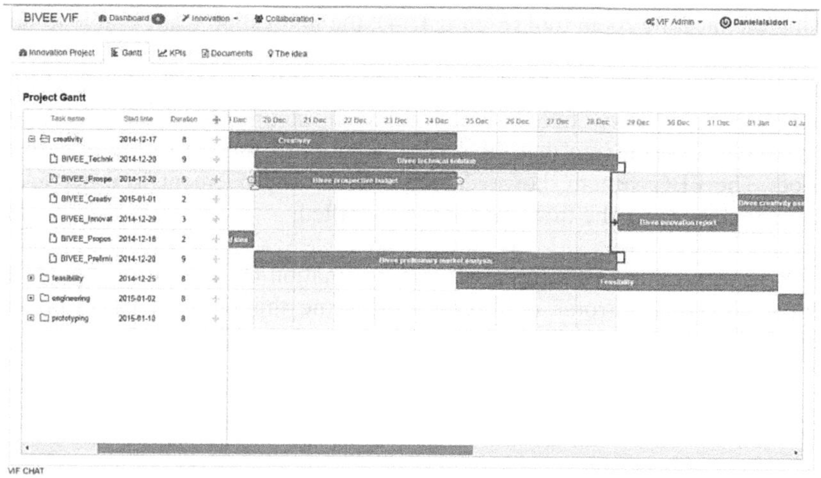

Figure 4.12. *Gantt tool in innovation project*

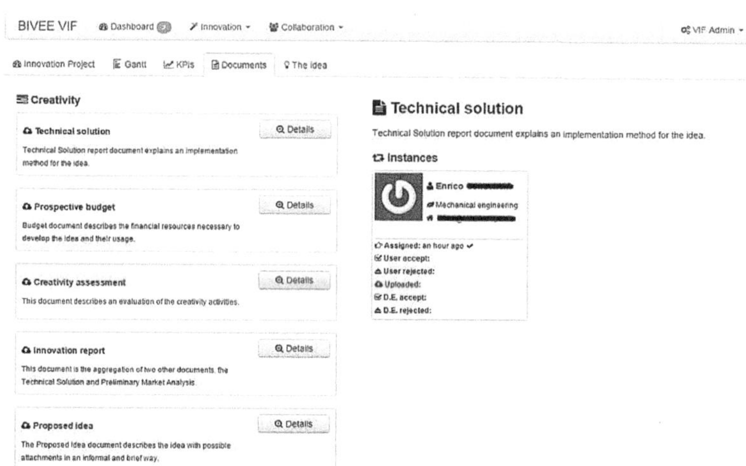

Figure 4.13. *Task management in Innovation Project*

The innovation team performs a study on a possible solution with feedback and comments using all the collaborative tools provided by VIF in order to collect more information. The VIF will take care of the constant assessment of the evolution of the project. After brainstorming and meetings using the meeting room tool (Figure 4.14), the innovation team is able to link the *proposed idea* to the *technical solution* because they decided to develop the solution based on the FBG system. The decision was supported by the fact that University tested both technical solutions (sonar and FBG technology), concluding that the second solution is much reliable in case of flood. The FBG system, however, requires an improvement in order to be engineered because the first version is too crafted.

It is very important to trace all the information collected in order not to repeat the same "errors" or discover the same things again based on the experience of other members of the VEE.

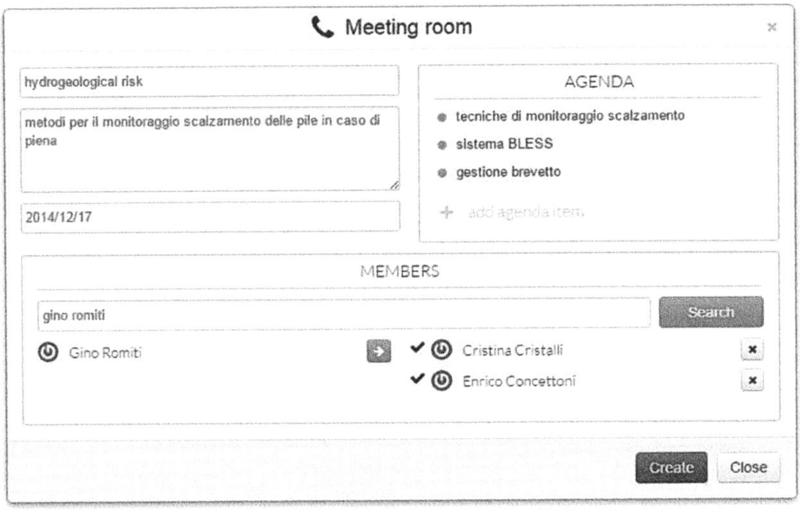

Figure 4.14. *Meeting room in collaboration tool*

Gino uploads the *marketing report* on the BIVEE platform and Cristina checks the state of the KPIs related to the *creativity wave* that are mainly related to time, innovative potential and costs. Then she uploads the *innovation report*, merging all the information from the *technical solution* and the *marketing reports* and sends it to Loccioni management.

The *innovation report* is the document that summarizes the contents related to technical solution and market report in order to propose the idea to Loccioni top management (Figure 4.15). They find the report very interesting and decide to proceed with the *feasibility wave* of the project.

Daniela reads the notice via mail and with much happiness writes and uploads the idea assessment after annotation (Figure 4.16). Now the gate is open and the project switches to *feasibility wave.*

Due to all the tools of VIF, the monitoring of KPIs and the availability of the documents the decision is simpler and the interaction among the members is very useful and immediate, so all the steps precede quickly and with an automatic tracking of all the interactions occurred.

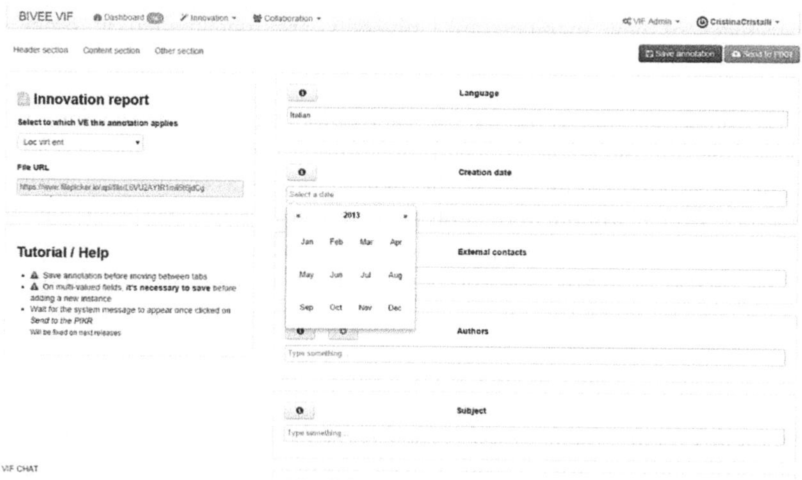

Figure 4.15. *Innovation report annotation*

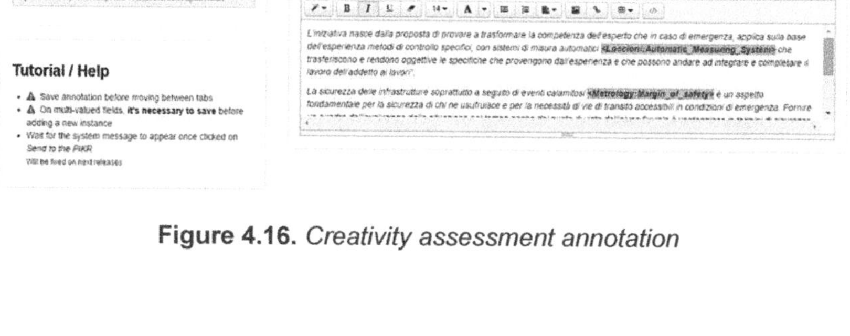

Figure 4.16. *Creativity assessment annotation*

4.4. Feasibility wave

The project now is entering the *feasibility wave*: from the idea we will start to develop and verify the assumptions taken during the *creativity wave.*

The Flumen project is interesting but it needs to be "challenged" and go through all the phases of the *feasibility wave* according to BIVEE – BIRF. Here researchers and business people work in cooperation to understand the technical and financial feasibility of the project. Furthermore, existing experiences of new members of University are fundamental in order to acquire further knowledge on the topic.

Technical solution proposed by innovation team has to be improved because the solution patented is still too crafted. In fact, the solution realized provides a heating circuit that in the first version is assembled manually to the Fiber Bragg without any protection device. In this new version, the innovation team wants to optimize the performance of device engineering in terms of sensor response, software and mechanical layout in order to have a performing and reliable system.

First of all, they have to choose the supplier for all the optical components. This is a very delicate aspect because the innovation team has to find the better solution in term of cost and performance. The KPI related to the number of new contacts of Flumen project (Figure 4.17) is automatically updated each time the team contacts a new supplier. The increasing of this KPI is a good aspect for the evaluation of the innovative potential of the project.

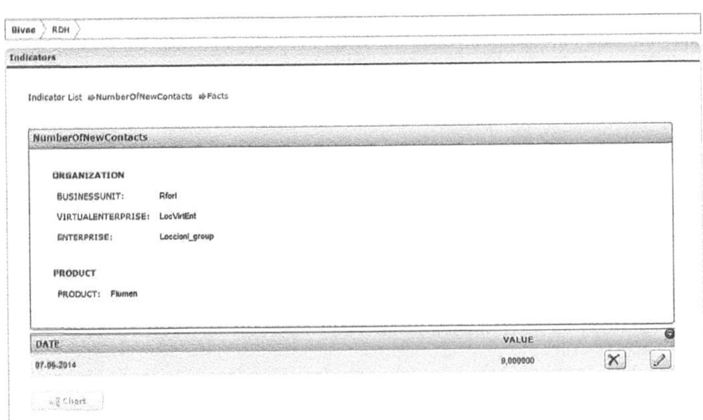

Figure 4.17. *KPI contact*

In the *feasibility project proposal*, Daniela explains the project, and the subprojects (in this case two: sensor assembly and installation) specifying again the members of innovation team and the required competencies. She uploads and annotates the documents and assigns the other task: Gino is again responsible for the *market analysis report* while Cristina uploads the *Alliance document* and all the members have to write the *periodical subproject reports* (Figure 4.18).

	BIVEE_Creativity_Assessment. Gate is opened to feasibility. Moving to next wave	
2 minutes ago	Cristina Cristalli assigned to document:BIVEE_Alliance	Daniela Isidori
a minute ago	Gino Romiti assigned to document:BIVEE_Market_Analysis	Daniela Isidori
a minute ago	Enrico Concettoni assigned to document:BIVEE_Periodical_Sub_Project_Report	Daniela Isidori
a few seconds ago	Daniela Isidori assigned to document:BIVEE_Project_Proposal	Daniela Isidori
a few seconds ago	Cristina Cristalli assigned to document:BIVEE_Feasibility_Assessment	Daniela Isidori

Figure 4.18. *Feasibility plan*

This phase appears to be redundant and time consuming, but it is necessary in order to optimize and minimize the risk of errors: technology and solution reuse is a very important aspect to take into consideration to keep costs low.

After several searches and conference calls using the meeting room, the innovation team finds a supplier who is an expert in the oil and gas industry who could answer their issue. In fact, the supplier can solve the problem related to the heating circuit with a more robust solution and he shares his solutions through the BIVEE platform. He promptly becomes a partner of the Loccioni VE and a member of the innovation team after Enrico's invitation.

Different components need to be tested before reaching a satisfactory solution. In fact, the innovation team decides to make a comparison between the solution already developed by the University and the one proposed by the supplier. In any case, the mechanical layout is implemented in order to obtain a much robust and portable system.

The KPIs monitored during the feasibility wave are related to:

– *cycle time, response time, reaction time, planned time and deviation from planned time*, in order to consider the time elapsed and the velocity of the innovation team;

– *costs, planned costs, deviation from planned costs and internal/external funding ratio*, in order to evaluate the economic sustainability of the project;

– *existing/needed knowledge ratio, number of new contacts, concept innovation index*, expected fulfillment ratio to consider the innovative potential.

Cristina prepares the *feasibility study report* and uploads it in the platform (together with the annotation synchronized with PIKR). In this document, the state of the art, the subprojects, and, very important, the risk analysis and the cost benefit analysis together with some recommendations for the realization phases are reported.

Having discussed the last technical aspects, the innovation team is ready to build the first prototype. In order to switch to the next wave, once again, Loccioni's top management has to evaluate the state of art of the project through the evaluation of KPIs and documents proposed. BIVEE platform allows checking together these two factors in a very useful and immediate way and this is fundamental in order to have a complete vision of the Flumen project.

Checking all the mentioned aspects, the management understands that this project has high risks, but they know that this is a very actual and crucial problem which can involve both public and private aspects.

In fact during the feasibility study, a big flood occurred and risked involving the headquarters of the Loccioni enterprise. Therefore this problem became of a high priority also because hydrogeological risk will represent a very important and actual national problem. For these reasons, Loccioni's top management decides to go on to the prototyping wave.

4.5. Prototyping wave

Having obtained the *feasibility study report*, the innovation team starts to build the first prototype. There are tight constraints on the budget while trying to maximize the reliability of the FBG system. Also in this case, Daniela assigns the tasks: the *BIVEE prototyping assessment* is assigned to Cristina, the *BIVEE implementation roadmap* and *BIVEE requirements* are written by Daniela and the *BIVEE prototype technical reports* are written by all the involved members. In this wave, it is normal to have several subprojects because in a complex and innovative project, there are a lot of parts to take care of. For this reason, a platform such as VIF is fundamental and very useful to manage all the steps, share knowledge and monitor the KPIs related to each step of the project. There are five subprojects in this case: the FBG system, civil project, electrical project, mechanical project and external heating circuit.

The first tests conducted in the laboratory prove that the solution proposed by the supplier is much more reliable with respect to the other. The team is satisfied with the new solution that is really robust, flexible and reliable. All the subproject reports have been completed in order to continue with the installation (Figure 4.19).

Figure 4.19. *3D model and prototype realized*

The *number of designs initiated KPI* indicates that several parts are produced and the *number of design changes KPI* indicates that all the designs have been changed (Figure 4.20). In fact, the Flumen project required a lot of efforts due to the high percentage of uncertainty and there are a lot of boundary conditions that have been changed during the installation.

The BIVEE environment plays an important role allowing a fast interaction among the members of the innovation team (both internal and external of the Loccioni group), in the organization of the reports and documents and in the analysis of KPIs.

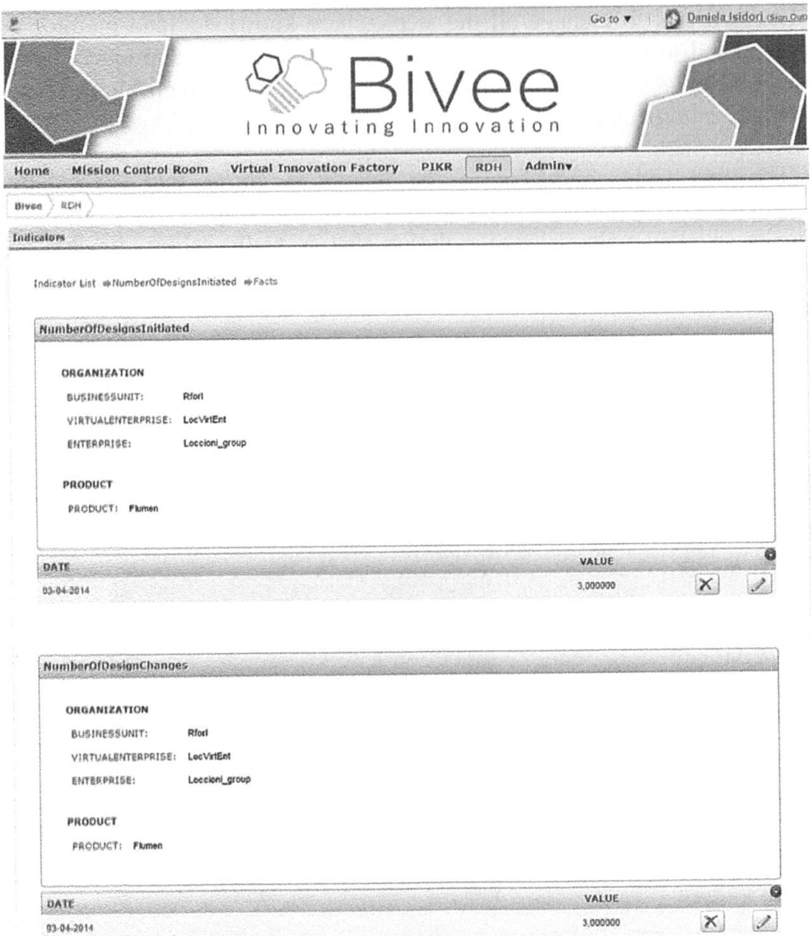

Figure 4.20. *Number of designs initiated and number of design changes KPIs*

Then, the system is installed on the Esino river bed, close to a pier of the bridge, which is considered the most exposed and vulnerable. During the installation, as already said, technical problems occurred and one of the two arrays installed was broken. Fortunately, the two arrays are installed for redundancy.

The deviation from planned cost assumes a positive value (Figure 4.21) but the results could be considered positive because the levels of risk and uncertainty were very high.

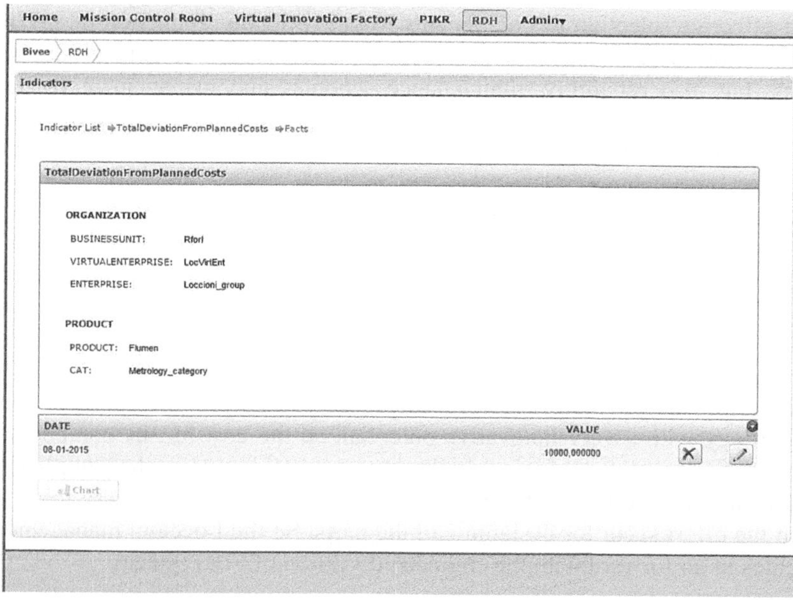

Figure 4.21. *Deviation from planned costs KPI*

There are many lessons learned and improvements that can be made on the layout of sensors and on the method used for the installation that have to be considered for the engineering wave. These achievements and feedback are reported on the *BIVEE Prototyping Assessment Report*. After a long and detailed measuring campaign, the innovation team establishes that the improvements suggested for the system make it ready for the engineering wave.

During the prototyping wave, the possibility to monitor in real-time the KPIs related to velocity and cost business objectives, together with the possibility to understand from all the reports which are the motivations that determined the choice of the last configuration of the system (rebuilding the "story of the project"), results in a very successful way to manage an innovation project.

4.6. Engineering wave

Having completed in a positive way the prototype, Loccioni's top management requires a careful analysis of all the components to guarantee cost-effective selection of components for a profitable production of several units of the system. Here, it is necessary to consider not only production costs, but also reliability, maintenance schemes and management of data obtained from sensors. These data will be used for managing hydrogeological risk and will be shared with public administration for the damage evaluation and alert management.

The results obtained from the prototyping wave show that from a technical point of view, the system works very well and it is consistent with the requirements previously requested, while from an economic point of view, the cost of the system is still too high with respect to the *expected commercial value KPI* (the two KPIs are shown in Figure 2.22). The innovation team is, however, confident that the FBG system is a new technology with a very innovative potential, so the cost for the components of this first prototype is too high, but acceptable. Also the supplier is confident with the fact that the costs for the materials will decrease together with the effort spent for designing all the parts. So the Loccioni management decides to go forward with the engineered version of FBG system.

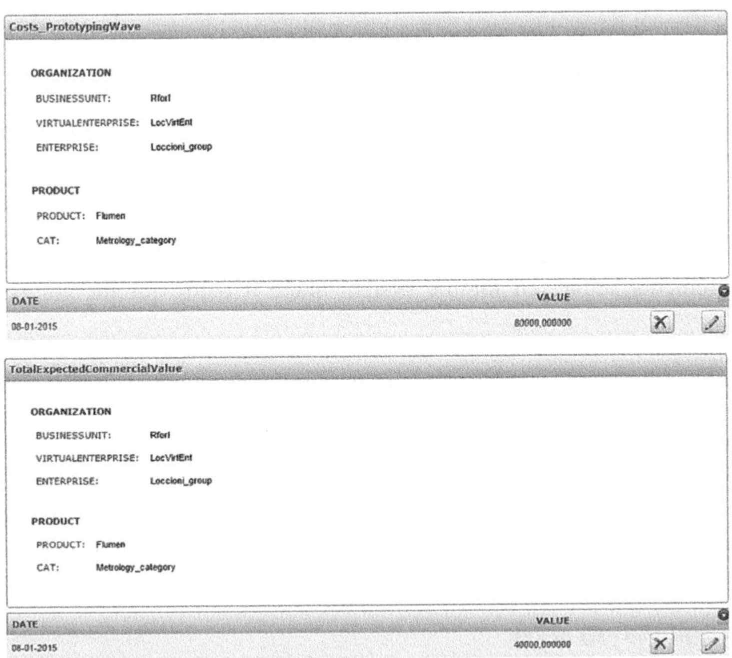

Figure 4.22. *Costs of prototyping wave KPI and total expected commercial value KPI*

At the moment the project is still in the *engineering wave*, waiting to realize the final version of the FBG system for a new costumers.

In this deliverable, two typical cases study of production and innovation cases are described. The two pilot cases are examples of the improvement of the two end users environments due to the use of BIVEE environment.

4.7. Conclusions

During the description of these pilot cases, all the improvements introduced by the BIVEE environment can be understood. Due to all the BIVEE tools, the monitoring of KPIs and the availability of schema and documents the decision is simpler and the interaction among the members of the VE is very useful and immediate. Through the BIVEE environment, all the steps quickly proceed with an automatic tracking of all the interactions occurred in innovation space.

From the end users' point of view, the BIVEE environment with its tools, and the BIVEE project in general, have been a useful approach and technological boost in order to accelerate the innovation process management in an innovative company.

5

A Day of Networked Production Improvement

This chapter describes the most relevant activities carried out in a networked production environment in the furniture sector. Although this is not a very widespread approach, it is becoming a popular approach which is being adopted by many small and medium-sized enterprises (SMEs) in the Spanish furniture sector.

A structure based on the different production phases (planning, sourcing, building and delivery) has been followed in order to give an organized vision about the production activities. The situation presented here corresponds to an ideal production scenario supported by business innovation in virtual enterprise environments (BIVEE) giving special relevance to those activities carried out on the ground to address the manufacturing of goods demanded in a particular order.

5.1. Resources involved

Here are listed the tools involved in the day-by-day activities of a furniture manufacturer which operates in the context of a networked production organization. This list includes a first section that reports traditional tools which are commonly used in the furniture sector to exchange communication between the different actors of the value chain. The second section lists the tools developed in BIVEE such as the key components adopted to achieve an integrated environment for all the actors of the

Chapter written by Fernando GIGANTE VALENCIA, Anil PACACI and Ali Anil SINACI.

production network. The third section of the list reports the specific actors involved in the trial.

Infrastructures:

– communication tools:

- phone,

- e-mails,

- other shared communication tools (wiki, chat rooms, skype, etc.);

– mobile apps.

Traditional communication means have been considered together with emerging technologies so, in the current production environments, both approaches coexist, particularly during the early adoption of any new change in the production management.

Software tools:

– Mission Control Room (MCR) [BIV 14a]:

- whiteboard,

- production map;

– Production and Innovation Knowledge Repository (PIKR):

- production knowledge repository,

- Key Performance Indicator (KPI) repository;

– existing software systems of end users:

- GdP[1] – Enterprise Resource Planning (ERP) developed under the consultancy of AIDIMA (Furniture, Wood and Packaging Technology Institute).

The MCR is one of the key components in the evolution of the production management illustrated in this chapter. This tool is used to define the production processes and the members involved in the production network. Furthermore, the assignment of resources to processes and the selection and monitoring of KPIs are also performed by using this software integrated into

1 http://www.aidima.es/gdp/.

the BIVEE environment. In particular, the KPI definition and monitoring is done through the communication between the MCR and PIKR, which stores the formal definition of the KPIs and manages the submitted values. However, the ERP deployed in the furniture company should be especially considered. In this case, the selected ERP is GdP – production management – developed under the consultancy of AIDIMA and focused on the needs of the furniture industry. In this context, the ERP is used to collect the KPI values from the real production environment in order to be managed by BIVEE.

Actors:

– sales department;

– university purchase department;

– technical office;

– logistics department;

– production department;

– carriers;

– suppliers.

The contact between the furniture company and the university is represented by the corresponding sales and purchase departments. This is an important issue before the starting of the production activities. The technical office of the furniture company reflects and formally defines any variation in the product properties (i.e. design and involved materials), raised from the negotiations between the aforementioned departments. Logistics and production departments are particularly involved in the production activities, while carriers and suppliers are considered external partners which play a key role in the networked production approach covered in this chapter.

5.2. Setting the scene

The furniture company, which is the main actor in this scenario, will be referred to as "the company" from now on. This is a local and traditional furniture manufacturer. The catalog of this company includes a wide range of products from bedroom suites to office furniture. To handle the complexity and variety of its products, the company does business with a lot of enterprises from different sectors such as board suppliers and shipping agencies. To handle the complexity of the production lifecycle, the company usually creates a temporary alliance with those aforementioned enterprises. This production network is called the *virtual enterprise*. To be able to better analyze and improve its processes, the company uses the BIVEE environment, mainly the MCR in addition to the ERP software.

One of the biggest universities in the country builds a new student dorm on one of its campuses. For these new rooms, 500 new wardrobes need to be purchased. After a quick search on the Internet, the purchase department of the university gets in contact with the sales representatives of the furniture company.

5.3. Plan phase

5.3.1. *Sales trend analysis*

For the company, the sales/trend analysis is an important activity for accurately analyzing the market. With this analysis, the company ensures an adequate manufacturing volume of furniture goods so it performs an accurate sales forecast. However, the demand can be estimated and a good stock level can be maintained in the warehouse in order to further facilitate the delivery of the incoming orders. Regarding the processed orders, BIVEE provides a graphical view for those indicators related to those sales aspects in order to facilitate performing a more accurate sales forecast.

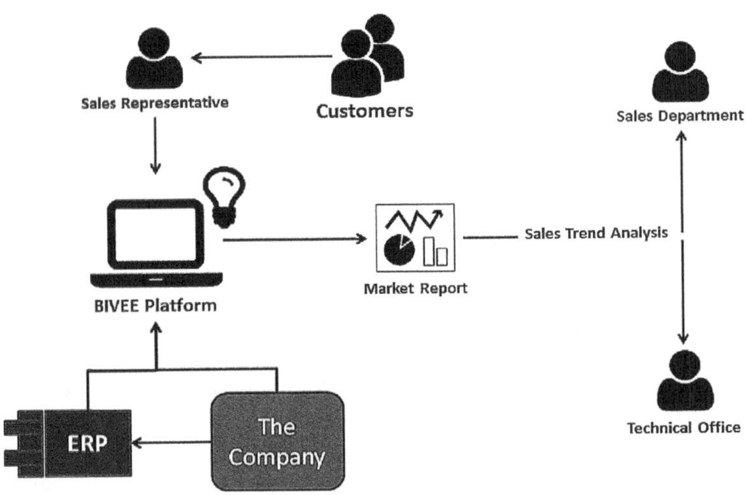

Figure 5.1. *Sales department making use of BIVEE*

The company has released a new two-door wardrobe to the market. To be able to maintain the adequate stock level, the sales department uses the historical information about launches of new products. As commented previously, this decision is supported by BIVEE and this is shown in Figure 5.1.

5.3.2. *Order evaluation*

The sales representative of the company visits the university campus to present the wardrobe products of the company and negotiates around a possible order. The management team of the university makes its choice on the recently released two-door wardrobe model. The managers state that the school season will start in two months and the order should be completely delivered in 50 days. The sales representative collects all the important aspects of the requested order so this could be properly evaluated by the sales department of the company. In order to formalize the process, the order request is sent to the sales department and the essentials of the order are registered and displayed on the BIVEE environment.

During this order evaluation, some critical aspects are identified. The volume of the order is relatively high and the delivery must be completed in just 50 days. In order to support the decision, within the BIVEE environment, KPIs obtained from former production projects (e.g. monthly number of processed orders and percentage of own transportation delivery orders) are examined. According to the evaluation, it is understood that the transportation of this amount of order to another part of the country cannot be covered by the internal delivery mechanisms of the company. The company aims to establish its network with the required supplier partners that this company mostly works with. Hence, a delivery partner will be required within the possible virtual enterprise in addition to the other production partners to fulfill this order.

5.3.3. *Product definition*

The technical office creates the product breakdown and defines the specific configuration of the requested product in the internal ERP. This document is the basis for deciding, during the source phase, the required materials, internal and external operations and services.

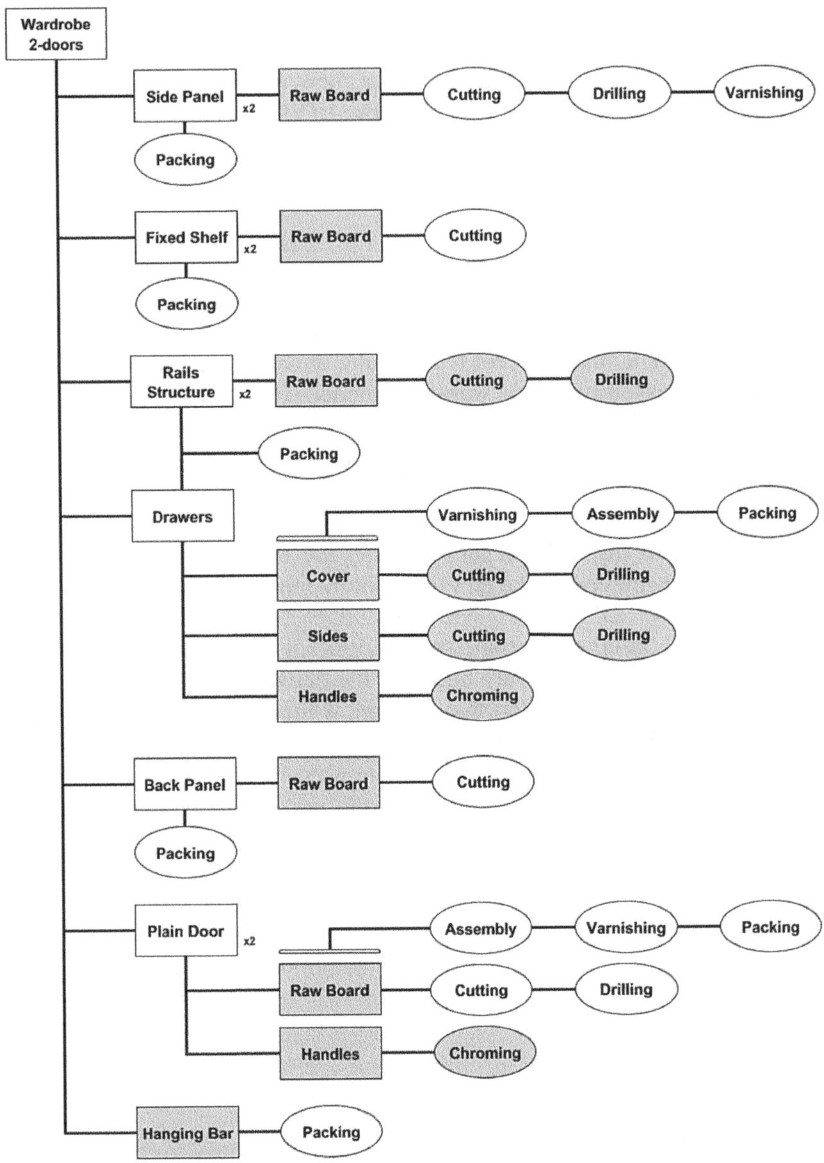

Figure 5.2. *Product breakdown of the requested wardrobes*

Figure 5.2 shows a simplified overview of the breakdown of the requested wardrobe. The boxes represent the materials and main components required for the manufacturing of one single wardrobe, while the ellipses represent the processes to be applied to the different materials and components. Those boxes in shaded background represent the materials and components that are needed to be provided by members of the virtual enterprise so the company is not able to produce them by its own resources given the order conditions. The same applies for the processes: those ellipses in shaded background represent the processes which should be performed by production partners in the virtual enterprise.

The product is modeled in BIVEE by using the MCR modeler. A new product structure model is created including the components required to produce the requested wardrobe, the used raw materials and those external processes to be applied to the product. Also, quantities and costs of each product part and processes are inserted into the product diagram. This model is an important resource for later evaluating the performance of the virtual enterprise.

5.3.4. *Network setup*

Those companies which are members of the virtual enterprise are introduced in the MCR using a scope model. Features for each company can be inserted such as contact information and production capacities. Companies in this model can be linked later to the processes required to fulfill the order.

By using the advanced search mechanism of the BIVEE environment, the business ecosystem is queried to find the most suitable partners for each specific operation: from the supply of raw materials to the completion of some external operations such as painting and varnishing. This search activity results in a number of candidate partners to fulfill each specific task in the defined virtual enterprise. Criteria such as cost and volume are introduced to obtain better results in a shorter period. The length of the configuration of the production network and its success are measured through KPIs such as “Cycle Time for Network Setup”.

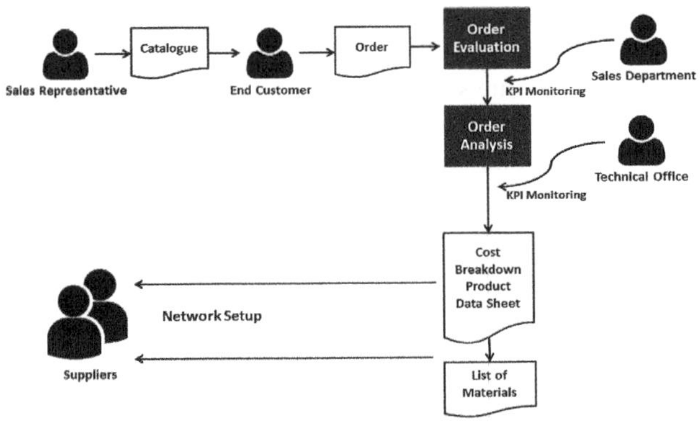

Figure 5.3. *Overview of the plan phase*

5.4. Source phase

5.4.1. *Stock analysis*

The technical office, the production and the logistics departments conduct a meeting to assess the feasibility of the new order request, considering the tight requirements. In this meeting, they examine the production breakdown document which is retrieved from BIVEE (as shown in Figure 5.3) and they decide if the order is manageable and if so, which production processes can be managed internally. During this activity, if the company has enough available stock in the warehouse of some external supplies, then the corresponding external producers are removed from the virtual enterprise.

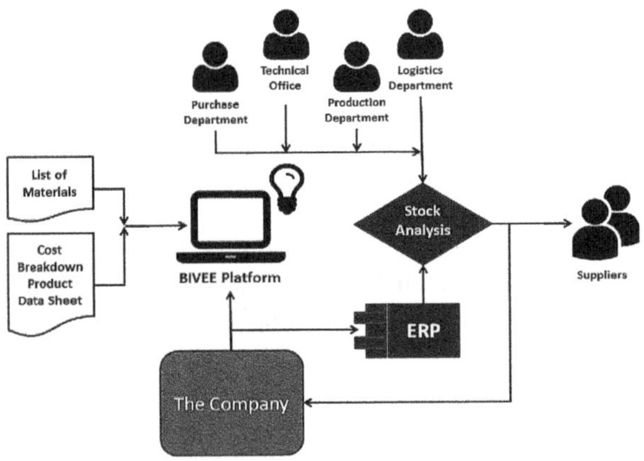

Figure 5.4. *Stock analysis with the support of the BIVEE environment*

5.4.2. *Supplier selection*

It is important to note that the company always wants to make business with the partners with which the company is used to making business. This is usual in traditional furniture SMEs and this has been especially taken into account.

During the meeting, materials and services which cannot be handled by the internal mechanisms of the company are identified. For these processes and services, external manufacturers and suppliers with required skills need to be found. Requests for quotations are sent to the partners selected with the search functionality provided by BIVEE through the MCR. All these requests are stored and managed in the BIVEE environment for future references.

During the preliminary evaluation of each production partner, the board supplier selected states that it is not able to deliver the required specified volume on time. The production plan is updated in the MCR and the production units posing problems are marked on the map. The technical office discusses possible solutions to this issue.

However, although prices are little bit higher, one of the delivery partners that the company mostly works with is chosen for this task by querying the BIVEE environment. The new carrier confirms the on-time delivery of the required materials.

Another process that should be managed externally is the handle manufacturing. Currently, there are two main handle manufacturers which have a close collaboration relationship with the company. The company HM-A produces high-quality and overpriced products, which decrease the profitability of the order. This issue can be measured through some KPI values (e.g. procurement price KPI). However, HM-B offers a lower price for low-quality handles which can cause incidents during the assembly process. This other factor can be measured through KPI values (e.g. incidents in assembly KPI).

To improve the overall production process, the technical office meets the innovation department to discuss possible improvements in the assembly process. It is decided to send feedbacks to HM-B about handle quality so that they can improve and change materials during the manufacturing of these components.

After deciding on each operation and its corresponding supplier, the production plan is instantiated from the product breakdown by the production department. The definitive virtual enterprise is created in the system. Each company of the Virtual Enterprise (VE) receives a notification and the registration status is delivered through BIVEE.

When all the invited enterprises join the VE in the BIVEE environment, the logistics department is notified on the creation of this virtual enterprise and inspects the production plan in order to validate it. After the finalization of the production plan, the members of the network are able to track the production activities having a big picture about the actors and processes involved and its specific responsibilities in the global production plan.

5.4.3. *Purchase management*

After receiving the improvement proposal from the company, the enterprise HM-B changes its production process using a different supplier for the raw material of the handles. This decision was taken based on the feedback given by the technical office of the company. By observing the KPIs displayed in the environment, personnel from the technical office again perform a quality-price trade-off analysis between both handle manufacturers. This is done to ensure that HM-B preferred over HM-A is able to meet the requirements and this will lead to a lower procurement price, checked by the inspection of the corresponding KPI value.

Sometimes, traditional manufacturers are reluctant to share their own KPI values with other companies. However, the access levels provided by the BIVEE environment enable a VE member to choose which information will be shared with whom. This is considered an improvement provided by BIVEE so it facilitates and promotes the information sharing among the members of the VE in this kind of production environments which are formed by traditional furniture SMEs.

With the new set of selected partners, the whole production process is simulated by using the MCR assistant of the mission control room. This allows testing the performance of the whole production network. After the validation process, it is detected that the proposed virtual enterprise is able to deliver the order on time. Given this, the university is notified about the acceptance of the order.

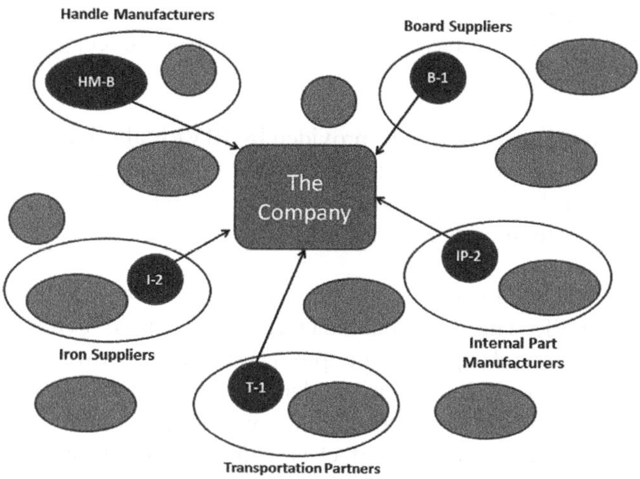

Figure 5.5. *Virtual enterprise formed over business ecosystem*

5.4.4. *Component storage*

This subphase is formed only by internal operations which remain in the scope of each internal company involved in the VE and in principle they are not eligible for being influenced by the rest of the VE members.

Despite this, this subphase is considered here due to the importance of the activities performed at this step. A correct performance in the activities performed at this point is especially relevant. In this regard, a wrong or ineffective warehouse management may lead to an important decrease in the performance of a specific production partner and this means a decrease in the performance of the VE as a whole.

5.5. Build phase

5.5.1. *Component manufacturing*

After the initialization of the production plan, the system provides a template list of production KPIs for each member of the VE. Each member is required to fill and update these values. Also, some KPIs are computed from exchanged documents and information provided by the existing software systems. This is the case of those KPIs related to delivery issues and customer satisfaction. GdP is able to manage some of these indicators automatically but others have to be collected from documents such as spreadsheets.

Document templates [BIV 14b] provided by BIVEE allow easy extraction of the information. In this way, the production processes can be monitored and each partner has a better vision of the VE through the monitoring facilities provided by the MCR assistant.

The production plan is updated by the manufacturers to inform the rest of members of the VE. The board manufacturer, for example, delivers the raw materials on time and updates the production plan. However, the iron supplier introduces a small delay in the delivery of the metallic pieces on the production plan. With this delay, the whole production process is automatically updated. Since the effect is negligible, an on-time delivery is predicted by the MCR of BIVEE after introducing these minor adjustments.

However, the company reports some problems on the handles provided by HM-B in the past. These are problems during the assembly of these pieces to the board doors of the wardrobe due to the quality of the chroming. So, the company has experience in this activity, this suggests a little change in the chroming process to the production partner in charge of applying this finishing. This is simply an increment in the time the metallic pieces are submerged in chrome in order to increase the thickness of the material applied. This way, the handles are more resistant and durable. HM-B evaluates the proposed change and accepts the new method. A new budget considering this new change in the production process is submitted to the company.

5.5.2. *Finishing*

In the proposed virtual enterprise, the finishing operations performed by the company remain out of the scope of the collaborative environment so they are carried out internally by the members of the production network. The given VE managed by the company delegates the required finishing operations to external partners. The chroming operation just mentioned previously is a clear example of this issue.

However, this subphase is included because it has important relevance in the product manufacturing and it provides a whole picture of the production cycle.

5.5.3. *Product assembly*

Given the properties of this kind of furniture, the whole product is not completely assembled during the production activities, so different parts are packaged and transported together. In this case, only some parts of the product are joined together (i.e. the components of the drawers).

In the meantime, a number of improvement ideas are proposed and rated by the employees. The innovation department considers and realizes one of these ideas trying to improve the assembly process with a new method which could involve some new equipment.

The idea is considered to be an improvement in the assembly process. This idea comes during the innovation activities carried out in the technical office which is in charge of the product development. Some prototypes are developed based on the new technique. The idea is based on the substitution of the gluing assembly with fittings and mounting hardware. This is a new way of assembly for the company and it has never been tested in the real production environment. This new technique of assembly is more expensive than the old technique but it presents many advantages: if some part is damaged after the final assembly of the product, the affected part can be replaced in an isolated way. Also, any incidence becomes much easier and cheaper so the damaged component can be replaced without affecting the whole product. Furthermore, the transportation becomes cheaper so the products can be folded reducing the space needed for the delivery.

In the light of these advantages, the production department decides to apply this new assembly process to the production of the wardrobes, observing a positive change on those KPI values related to the corresponding activities.

5.5.4. *Quality control*

Once all the external processes are completed and the products are assembled without any delay, the production status is updated on the BIVEE environment.

In the internal scope of each company of the virtual enterprise, some quality control specifications are managed in order to formalize the product checking process. The company, which plays the role of administrator of the VE, decides to attach its quality control specifications to the system in order to share it with the rest of the VE members and also with other colleagues inside the company.

Through the monitoring facility and KPIs, each company can better analyze the processes and the products assembled. Changes introduced to the production plan affect in particular the "Cycle time of assembly" KPI and the "Incidents in assembly" KPI. The technical office pays more attention to these KPIs in order to understand whether the changes result in some improvement or not. In this regard, it is observed that both KPIs reveal an overall improvement.

5.6. Delivery phase

5.6.1. *Packing*

As soon as the quality control finishes and all the products are ready to be delivered, the packing activity starts. During this process, the produced wardrobes are enclosed for storage and later delivery. Goods are packaged according to the specifications defined based on the type of product, in order to protect the pieces from possible damages during the transportation phase. However, packages are labeled with the internal code of each packaged product.

Activities performed during this subphase remain in the internal scope of the different VE members. At this point, the delivery phase formally starts.

5.6.2. *Order preparation*

At this step, packages related to the order are collected. Due to the specific configuration of the requested products, the handles are placed according to the design of the selected configuration of the wardrobes. Also, the fittings required to install the products in the destination place are prepared during this step.

When these activities have finished, the transportation partner is notified about the prepared orders. However, the partner updates its status on BIVEE and states that it is not possible to manage the delivery due to unexpected truck breakdowns.

The only option to deliver the order on time is a partner replacement. The logistics department swiftly performs a search using BIVEE through the business ecosystem (Figure 5.5) and finds an alternative partner for the transportation. During this activity, the delivery notes are generated according to the information stored in the BIVEE environment.

5.6.3. *Shipping*

When the virtual enterprise is updated with the new member, the shipping activities start with the labeling of the packages containing the wardrobes. Packages are labeled with the code of the order. The carrier name and the final destination of the produced goods, in this case the address of the university campus, are indicated in these labels.

The delivery orders previously generated are accessible to the carriers at any moment through the BIVEE environment. Due to these documents, the transportation process can be observed by all the partners, in particular by the company which is the VE administrator. Once the goods are properly loaded, a notification is sent to the company to keep it informed about the status of the delivery process, indicating that the goods are already prepared in order to be delivered.

5.6.4. *Delivery*

Once the products are delivered to the university facilities, a notification is sent to the company to keep it again informed about the status of the order. After this, values of the production KPIs are analyzed through the MCR of BIVEE.

Any possible achievement and improvement identified regarding the production process is reported to the research and innovation department through the BIVEE environment. BIVEE populates the information available in the environment automatically, such as KPI values. This means a reduction of the time required to send the report.

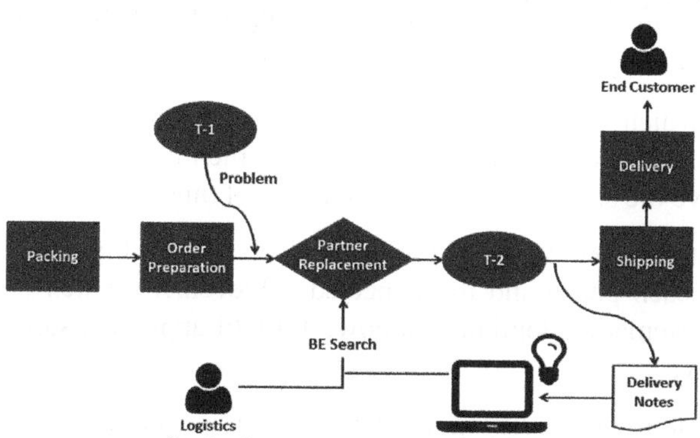

Figure 5.6. *Overview of the delivery phase*

During this phase, some important documents are managed, such as the packing instructions, packing list, delivery notes and package labels. These documents are annotated and managed by the BIVEE, improving the exchange of all the relevant information among the involved actors.

At this step, the company is able to evaluate the performance of the virtual enterprise so the BIVEE has kept track of all the relevant indicators involved in the production phases. In this regard, those indicators related to the customer satisfaction business objective are considered the most important. This satisfaction can be measured through the values of KPIs such as (1) quality level of executed orders, (2) percentage of product damage, (3) percentage of returned orders and (4) percentage of claims.

Furthermore, KPIs such as “Total production cycle time” and “Total production price” can be better evaluated at the end of the production cycle.

5.7. Final considerations

The whole production process, through the four BIVEE production phases [BIV 14c] (plan, source, build and delivery), is fully supported by the BIVEE environment, with:

– all the possible communication facilities intra- and interworking teams;

– advanced search features with criteria support on business ecosystem to determine the most relevant and appropriate partners;

– support to the ongoing work, having defined a number of document templates which clearly specifies all required information and steps to follow to ensure the quality of the information collected and the accuracy of the dissemination of the information within the business ecosystem;

– MCR which enables us to simulate production processes during the source phase and provides a production plan which helps the members of the VE to observe the status of the production in real time;

– a constant monitoring of the time, cost and quality of the outcomes returned by each phase and the expected achievement which is carried out due to the systematic adoption of a powerful KPI approach, supported by the KPI ontology;

– integration with existing software, such as the ERP to enable the automatic extraction of KPI values to enable the monitoring of the performance of the virtual enterprise.

5.8. Bibliography

[BIV 14a] BIVEE deliverable 3.22 – Mission Control Room Prototype v2.0, available at http://wordpress.bivee.eu/wp-content/uploads/2014/11/D3.22-Mission-Control-Room-Prototype-v2.0.pdf, 2014.

[BIV 14b] Technical annex for BIVEE deliverable D6.12 – end-user analysis through document centric approach, available at http://wordpress.bivee.eu/wp-content/uploads/2014/10/D6.12-End-User-Analysis_Annex.pdf, 2014.

[BIV 14c] BIVEE deliverable 2.2 – specification of business innovation reference framework, available at http://wordpress.bivee.eu/wp-content/uploads/2014/10/D2.2-Specification-of-business-innovation-reference.pdf, 2014.

PART 3

Innovating Innovation: BIVEE Achievements

6

The BIVEE Framework and the Collaborative Innovation Capability Maturity Model (CICMM)

Similar to most software development projects, the business innovation in virtual enterprise environment (BIVEE) solutions are also based on a framework to provide a guiding structure. The BIVEE framework is described within this chapter, starting with its objectives and methodological rooting. Additionally, the concept of the collaborative innovation capability maturity model (CICMM) is described, which is a model for the assessment of a network's management maturity and conformity concerning collaborative innovation management.

One main notion of the BIVEE project is the differentiation between the innovation space and the production space of virtual enterprises. Accordingly, the BIVEE framework supports this approach in order to provide specific structure for both parts of the collaboration. The BIVEE framework differs between improvement processes that mainly appear within the production space and innovation projects emerging from the business space. Due to the addition of a separate monitoring framework (MF), the BIVEE framework is composed of three subframeworks (Figure 6.1):

– The construction of the BIVEE framework is based on the *virtual enterprise modeling framework (VEMF)*. It contains a framework for the general setup of the virtual enterprise, and defines its production processes by providing a modeling framework for the general structure of operative business.

Chapter written by Benjamin KNOKE.

– The second part of the BIVEE framework is the *business innovation reference framework (BIRF)*. It is called a reference as it does not enforce a strict structure, but a guiding reference to innovation projects, because the latter usually does not progress in defined ways, especially during the early stages of idea creation. Additional to a reference of operational activities, the BIRF adds characteristics of four areas of innovation by defining templates for specific information input and output mapping related to each activity.

– The third component is provided by the *MF*, which adds metrics that are aligned to business (Table 6.5) or sustainability objectives (Table 6.6). These metrics are key performance indicators (KPIs) that either address the production (P-KPIs) or refer to innovation activities (I-KPIs).

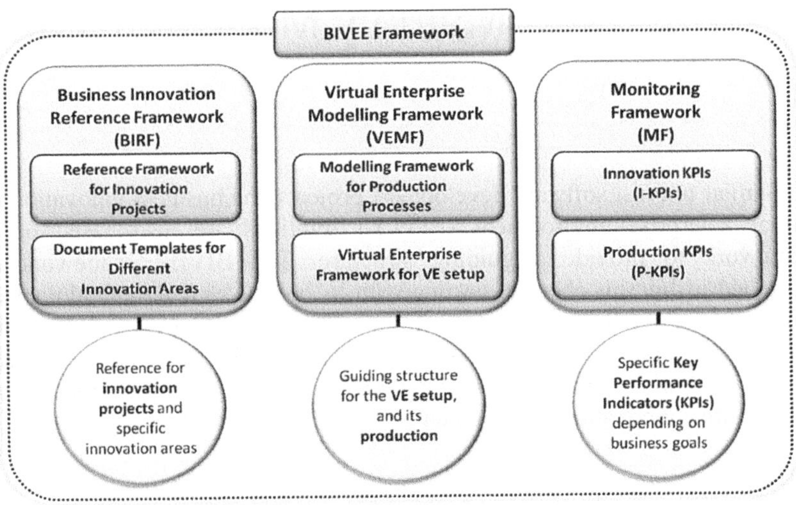

Figure 6.1. *Overview of the BIVEE framework components*

In addition to the three subframeworks and their methodological rooting, the concept of the CICMM is also described in the following sections.

6.1. The virtual enterprise modeling framework (VEMF)

The VEMF has been designed to provide a guiding structure during the setup of a virtual enterprise and its production processes.

6.1.1. *VEMF: methodological background*

The structure for the VEMF is mainly based on the value reference model (VRM) and the supply chain operations reference (SCOR).

The VRM is an open source dictionary developed by the value chain group and accordingly focuses on value chains. Because value chains represent a specific, although static, form of market-oriented collaboration, it has been possible to adapt the VRM to virtual enterprises in order to build upon a proven methodology. The structure of the VRM model supports and enables corporations to integrate their four critical domains: enterprise management, product development, supply chain integration and customer relationship management, using one reference model to support the vision of an integrated value chain [KIR 11]. Key elements of the standard process VRM dictionary include inputs/outputs, metrics and best practices. Processes described by VRM differentiate in three hierarchical levels:

– *Govern*: this level is composed of decision-based processes that identify and enable a value chain by establishing the rules, policies and procedures to control the implementation of plan and execute processes.

– *Plan*: this level is composed of processes which meet the current strategic objectives to current resources, and produce decisions on activities to reach organizational goals.

– *Execute*: this level is composed of processes that concern transformation of work and material to product or services to meet customer requirements. The processes in execute level operate within the limits of management criteria and parameters defined by the processes in plan level.

For the setup of the virtual enterprise, the processes concerning the value chain setup of the plan and govern sections are relevant. The other elements and processes of the execute section are replaced by the content of the BIRF and the production model of the VEMF, which provide structure for the operative business.

A second influence to the VEMF arises from the SCOR. SCOR is a product of the supply chain council (SCC) and serves as a process reference that proposes standard guidelines for supply chain management [STE 01]. The framework is built around two basic sets: a standard library for the modeling of supply chain processes, which allows companies to set up environments for collaboration and process alignment, and a set of standardized semantically defined metrics, which allows supply chain managers to quantify and benchmark processes and thus to drive supply chain management strategies.

The SCOR process map is structured into five recurring phases:

– *Plan*: this stage concerns developing plans to operate supply chain. This process describes aggregation and prioritization of requirements, inventory and capacity planning, and projection of appropriate resources to meet requirements of supply chain for long time period.

– *Source*: this process concerns scheduling of deliveries, reception of goods and services, authorization payment for materials and other purchases.

– *Make*: this process concerns conversion of materials or creation of the content of services. It is starting with requests for materials – linking to source process – and ending with product release linking to deliver process.

– *Deliver*: this process concerns the creation, maintenance and fulfillment of customer orders, scheduling order delivery, packing and shipment and invoicing customer.

– *Return*: this process concerns reverse flow of goods. The return process is composed of identification of the need to return, the disposition decision-making, the scheduling of the return and the shipment and receipt of the returned goods.

6.1.2. *VEMF: virtual enterprise setup*

The VRM represents a reference framework that provides a guideline for the management of value chains. As value chains can be characterized as a specific type of network, some elements apply to virtual enterprises as well. In particular, the govern section that is describing that the setup of the value chain can be used to support the virtual enterprise setup.

The section "Govern Value Chain" of the VRM resembles the setup phase of a value chain that defines a common strategy and vision as well as necessary policies concerning all common fields of action. This set of processes has to be executed by the initiating partners of a virtual enterprise as well. The policies and the prioritization of business objectives that are defined here are necessary for further progression. Based on this section, the "Govern Virtual Enterprise" section of the VEMF can be created (Table 6.1).

Name	Description
GVE01 – Rules & Objectives	To develop the strategy, vision, goals and business objectives for the virtual enterprise
GVE02 – Process Policy	To establish the process performance criteria, priority dimensions and supporting metrics
GVE03 – Information Policy	To define the creation, management, storage, distribution, retrieval and disposal of information in the virtual enterprise
GVE04 – Financial Policy	To define the accounting, financial decision support, routing and reporting for the virtual enterprise
GVE05 – Asset Policy	To define the use of intellectual, brand, physical, human and working capital of the virtual enterprise
GVE06 – Organizational Policy	To define the way the virtual enterprise partners will interact with the stakeholders as a corporate citizen to comply with regulation
GVE07 – Extended Network Policy	To define rules and policies governing partner interaction
GVE08 – Change Guidelines	To define how the virtual enterprise will deal with changing external/internal factors and evolve to meet stakeholder expectations
GVE09 – Compliance Policy	To define how the virtual enterprise partners will comply with regulations and legal requirements
GVE10 – Lifecycle Policy	Define lifecycle policies of a virtual enterprise

Table 6.1. *VEMF govern processes*

The govern processes are executed by an initial Virtual Enterprise (VE) consortium in order to define the rules and objectives as well as the policies regarding all important areas of business. This part provides the definition and prioritization of business objectives (GVE01) that are important within the monitoring framework (MF) as well as policies that impact the exchanged information within the business innovation reference framework (BIRF).

The definition of the network structure is a recurring process, as the need for resources and competencies may change during operation. Within the VRM, this part is resembled by the plan value chain section. This plan section is implemented into the VEMF with the plan virtual enterprise section, as shown in Table 6.2. It resembles the process of matching the virtual enterprises' resources to its needs by gathering the requirements, assessing and aligning resources and the creation of a business plan. The planning processes are executed whenever the needs of the virtual enterprises change. This mapping is checked and used within activities of both spaces.

Name	Description
PVE01 – Gather VE Requirements	The process of gathering virtual enterprise requirements necessary to fulfill market demand for comparison against available resources
PVE02 – Assess VE Resources	The process of assessing virtual enterprise resources for alignment with requirements needed to fulfill market demands
PVE03 – Align VE Resources	The process of aligning demand with required resources to capitalize on market opportunities
PVE04 – Create VE Business Plan	The establishment and communication of plans and actions over appropriate time horizons representing a projected appropriation of virtual enterprise resources to meet market requirements

Table 6.2. *VEMF plan processes*

These govern and plan sections represent the virtual enterprise framework. Other parts of the VRM are either not practical or are covered by other parts of the BIVEE framework. The production space processes are aligned to the operational level and replace the execute section of the VRM. These processes are described in the following section.

6.1.3. *VEMF: modeling framework for production processes*

The production process modeling framework of the VEMF is based on the SCOR processes. The processes of the SCOR model have been adopted, as well as regrouped in order to account for the distributed structure of VEs and the separation of business from production. The VEMF encompasses all activities related to the production spaces toward the goals of virtual enterprise. The four resulting phases are plan, source, build and deliver, which are described below:

– *Plan*: in this process level, requirement results of the global competition, the requirements of customers, stockholders, workforce, suppliers, as well as economic, technological, social, legal and political environments are taken as inputs, and as outputs, key performance objectives of virtual enterprise are defined and design decisions, such as business models and a business strategy map, are created. Furthermore, according to strategic decisions, products and services are designed, and detail production processes and appropriate resources meeting production processes for each production unit (PU) in value production space (VPS) in virtual enterprise are defined.

– *Source*: after products and services are designed and production processes are defined, appropriate resources are defined in this phase allocation, provided that deliveries and logistic works of resources are considered.

– *Build*: as products and services are designed, production processes are defined and appropriate resources are provided, in this process level, conversions of materials and immaterial into product or creation of the content of services with execution of production process that occurs in each PU take place.

– *Deliver*: this process level concerns order management, pricing and product configuration through managing orders, invoicing and credits, to customer-specific packing, labeling and shipping.

Additionally, the *design and monitor* process is defined. This process is not aligned to a specific phase, but instead spans across plan, source, build and deliver as a cross-cutting action. The related metrics are provided with the key performance indicators (KPIs) of the MF. An overview of the phases and their processes along with a short description is listed in Table 6.3.

Phases	Processes	Short description
Plan phase	VPS-P01 Sales Trend Analysis	Results of an initial market analysis show demands for production
	VPS-P02 Order Evaluation	An order may initiate a production process – needs to be evaluated
	VPS-P03 Product Definition	Order analysis and product allocation
	VPS-P04 Network Setup	Creation of a network to fulfill production demands
Source phase	VPS-S01 Stock Analysis	Analysis of individual stocks for purchase management
	VPS-S02 Supplier Selection	Selection of suppliers for raw materials
	VPS-S03 Purchase Management	Allocation of resources for production
	VPS-S04 Component Storage	Storage of allocated resources
Build phase	VPS-B01 Component Manufacturing	Manufacturing of components
	VPS-B02 Finishing	Final post processing of the production
	VPS-B03 Production Assembly	Assembly of components
	VPS-B04 Quality Control	Quality checks of manufactured goods
Deliver phase	VPS-D01 Packing	Packaging and preparing for delivery
	VPS-D02 Order Preparation	Commissioning of ordered products
	VPS-D03 Shipping	Carrier selection and shipping of manufactured goods
	VPS-D04 Delivery	Final release and delivery of manufactured goods

Table 6.3. *Production space processes and phases*

6.2. Business innovation reference framework

The BIRF aims to support collaborative innovation management by providing a structure for the necessary actions and the exchanged information depending on the innovation area.

6.2.1. *BIRF: methodological background*

The BIVEE approach with respect to innovation tries to find a compromise between guidance and freedom; since the latter is the natural nurturing ground for innovation but in absence of the former endless loops may be encountered that are very risky in a business context. So, in order to give some guidance to the innovation space, the BIRF is organized into four main waves that are based on the findings of the LABORANOVA Project [THO 07]:

– *Creativity wave*: the first wave starts with an innovation idea or a problem to be solved, providing a first sketchy idea to be developed and put into the business context of the virtual enterprise undertaking it.

– *Feasibility wave*: this second wave provides and clarifies with elements and information in order to justify in economical and operational terms the actual undertaking and the further development of the original idea.

– *Prototyping wave*: this wave puts into concrete actions the previous collected information and elements producing a first implementation of the initial idea in the form of a prototype.

– *Engineering wave*: in this wave, the original idea transformed into a prototype is attentively analyzed to generate production and engineering plans.

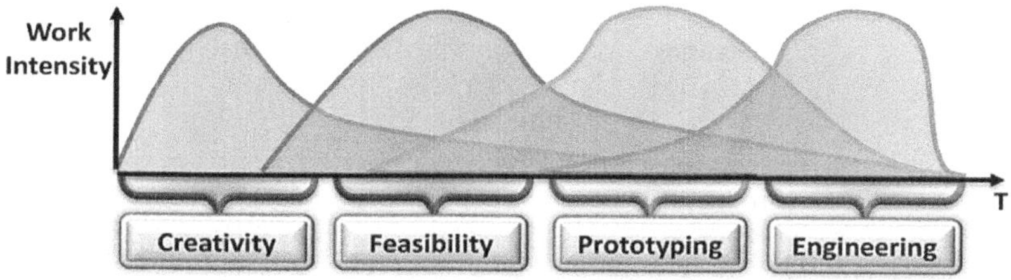

Figure 6.2. *The four BIVEE waves of innovation within the innovation space*

The four waves can be represented as shown in Figure 6.2. The innovation waves are not in a strict sequence, they are partially overlapped and they develop in cycles. We can assume that, for example, during the prototyping wave there may be a problem that requires a focused activity to look back into the creativity wave. A goal-decomposition approach is, therefore, extensively adopted in the BIRF, so there may be activities in each wave that proceed faster and can lead to partial prototyping while other activities that are still in the feasibility wave. This approach gives more flexibility and freedom to the innovation space. However, what guarantees a certain order is the concept of gates or innovation gates. At a given point of a wave, each project could set its own gate from which the activities of the next wave could start to activate. This notion is designed in the style of the Stage-Gate model [COO 02]. However, given the nature of the waves, the gates are not then a fixed point but can differ from project to project. Once again, this should ensure flexibility in the innovation cycle for each innovation project and for each virtual enterprise that is created *ad hoc* for that defined project, but, at the same time, should ensure the flowing of cycle itself.

The intention of this part of the BIRF is to include the differences between certain types of innovation in the BIVEE framework. Therefore, it differs between four areas of innovation. These areas are based on the approach of systematic business innovation by Kotelnikov [KOT 08], as shown in Figure 6.3. Regarding the more technical view of the BIVEE project, the areas of product, process, service and technology innovation are included. The differences between these areas are characterized mainly by the information that is processed within these activities and is resembled by the input/output mapping of this subframework.

Figure 6.3. *The four BIVEE waves of innovation within the innovation space [KOT 08]*

6.2.2. *BIRF: reference framework for innovation projects*

As previously described, the BIVEE project imagines the innovation activities to be divided into four waves instead of the more classical and linear phases. Each wave contains a certain number of activities to be carried out in order for a wave to be completed. These activities have been identified through a combination of the VRM, the findings from the LABORANOVA project and analysis of the BIVEE end-user's innovation practices. The resulting activities are listed in Table 6.4.

Waves	Activities	Description
Creativity wave	BIS-C01 Idea & Solution Generation	This initial activity produces an idea
	BIS-C02 Idea & Solution Analysis	First check of the idea
	BIS-C03 Initial Market Analysis	First market check and approximation of expected turnover
	BIS-C04 Final Solution Assessment	Decision to proceed or not with the idea validation
Feasibility wave	BIS-F01 Resource Analysis	Analyzing and comparing the available and needed resources
	BIS-F02 Feasibility Study	Creating a feasibility study of the concept for further processing
	BIS-F03 Proof of Concept	Approval or rejection of the concept
Prototyping wave	BIS-P01 Resource Allocation	Allocation of resources for prototype design
	BIS-P02 Design Prototype	Designing of the prototype
	BIS-P03 Validate Prototype	Validation and improvement of the design
Engineering wave	BIS-E01 Acquire Resources	Allocation of resources for final engineering
	BIS-E02 Build and Optimize	Building and optimizing the solution
	BIS-E03 Launch Planning	Marketing, communication and sales activities to support the launch of the solution
	BIS-E04 Release	Final process to support the production of the solution

Table 6.4. *Innovation space activities and waves*

Each of the activities is linked to an individual set of KPIs and information artifacts. The KPIs are part of the MF and the information artifacts are also dependent on the area of innovation – as described within the following section.

The first part of the BIRF contains a set of activities. These activities process information that either applies for all areas of innovation and improvement or is characteristic for only a few or a single area.

The BIVEE approach is “document-based”. An innovation project needs to walk through the waves and its activities before entering the production space. It does so by producing certain documents for each wave and more specifically for each activity of each wave. For each activity and for each wave, therefore, a set of documents can be defined that needs to be followed and completed in order to start the following activity and/or the following wave. These document templates can be found in the production and innovation knowledge repository (PIKR).

The innovation documents can be accomplished in any possible sequence, as long as some precedence constraints are satisfied (e.g. cost analysis cannot be performed before knowing the composition of a product). Each document has quality check criteria associated, to be used (together with the phase specific criteria) in checking if a phase has been accomplished.

The document templates selected from the PIKR are then progressively instantiated, while the innovation project proceeds. However, the sequencing is quite flexible and new documents could be added if needed. Furthermore, it is possible to move forward and backward across phases, but at the given moment each phase should be accomplished according to the defined criteria (local to each selected doc and global to each phase) and certified by an assessment document.

6.3. Monitoring framework

The MF aims to make the innovation activities and production processes manageable. Therefore, two sets of KPIs have been developed in order to monitor the activities within the two spaces.

The KPIs are aligned to a set of business objectives that have been developed according to the VRM and in close cooperation with the end-users of the BIVEE project. These objectives are identical for both spaces and listed in Table 6.5.

Name	Description
Reliability	The objective of constantly achieving operative goals
Velocity	The objective of fast reaction and short cycle times
Adaptability	The objective of being able to adapt to varying external or internal demands
Cost orientation	The objective of operating on minimal costs
Asset orientation	The objective of generating tangible and intangible profit
Innovative potential	The objective of successfully carrying new ideas to the market
Customer orientation	The objective of satisfying customers
Network orientation	The objective of a maximal efficient network

Table 6.5. *Business objectives of the monitoring framework*

BIVEE addresses manufacturers, who usually operate within socio-technical systems. The sustainability of these systems is usually dependent on economic, social and ecological dimensions [HOL 01]. Therefore, social and ecological aspects are added. The social performance of a system is defined by its ability to fulfill the needs of its stakeholders. The satisfaction of these needs is a complex topic that expands into multiple categories [DIE 97, GEI 06]. Social performance determines the dedication of a system toward society and ethical business. Categories of social performance have been refined from the work of [VAL 12] and [EDU 09]. These categories are employment, well-being, skill development, community interaction and accountability.

Due to the subjective nature of many indicators within the social performance context, their related measurements cannot be taken automatically by sensors. They are not based on artificial system properties, but instead on individual opinions of human beings. Therefore, in most of the cases, the performance within these categories can only be measured indirectly by interpretation of causal effects (e.g. ratio of staff away sick) or by conduction of surveys among the system's stakeholders. As a third dimension of sustainability, the ecological performance of VEs can also be assessed by certain indicators. The ecological performance comprises all interactions of the VE with its natural environment, which can be sorted into ecological input and ecological output. A brief description of the sustainability objectives is provided Table 6.6.

Name	Description
Employment	The objective of social benefit in the contractual relationship between an organization and its staff
Well-being	The objective of satisfaction of employees
Skill development	The objective of providing options for training and skill enhancement for employees
Community interaction	The objective of quality and quantity in social interactions with the environment
Accountability	The objective of an organization to take responsibility for its actions
Ecological input	The objective of achieving an environmentally friendly use of resources of a system
Ecological output	The objective of achieving an environmentally friendly emission of a system

Table 6.6. *Sustainability objectives of the monitoring framework*

The performance of a VE within these objectives can be measured through its impacts on environmental factors, e.g. which may concern global warming, acidification, toxic contamination, biodiversity or radiation. Such environmental themes have been taken from [WAL 00] and create the basis for the design of ecological KPIs. The Canadian Environmental Sustainability Indicators (CESIs) are an example of the transparent publication of an organization's ecological KPIs, whereby the organization in this case is Canada itself.

Based on the objectives described in Tables 6.5 and 6.6, a set of reference KPIs has been developed in interaction with the BIVEE end-users. The MF currently holds a total of 69 KPIs for the production space, 100 KPIs for the innovation space and 29 KPIs concerning the sustainability context in both spaces. The selection of the KPIs that are communicated globally within the VE is made during the govern process (GVE02 – Process Policy).

6.4. Collaborative innovation capability maturity model (CICMM)

Through governing processes, collaborating organizations should aim for a conjoint set of KPIs and a high level of management maturity. To guide both, the improvement of innovation actions and the consolidation of the network's methodology, the concept of CICMM can be applied [KNO 13]. This concept is based on the capability maturity model (CMM) [CMM 02]. The CMM was initially developed to increase the maturity of software development processes, but can also be applied on innovation management [ESS 10]. The following five levels of innovation capability maturity are used within the CICMM:

– *initial*: the processes are unpredictable, poorly controlled and reactive;

– *managed*: the processes are characterized for projects and are often reactive;

– *defined*: the processes are characterized for the organization and are proactive;

– *quantitatively managed*: the processes are measured and controlled;

– *optimized*: focus on process improvement.

In addition to the maturity monitoring, a second dimension is added to account for the consolidation of the network's approach on innovation management. This notion of network conformity consists of four additional levels:

– *Single organization*: the approach is realized in a smaller part of the network. Practices of other partners in this area are unknown.

– *Network awareness*: the approach is implemented by a single or a few partners. Other partners are informed about this best practice and invited to join.

– *Network consent*: all relevant innovators within the network agree to this approach, but it is not working or used properly.

– *Network dedication*: the lowest level of innovation capability maturity that has been gained by all relevant innovators within the network.

The data to assess a network's maturity and conformity can be gathered either through observation or a Web-based questionnaire. Once collected, the assessment can be visualized in a two-dimensional chart, as shown in Figure 6.4.

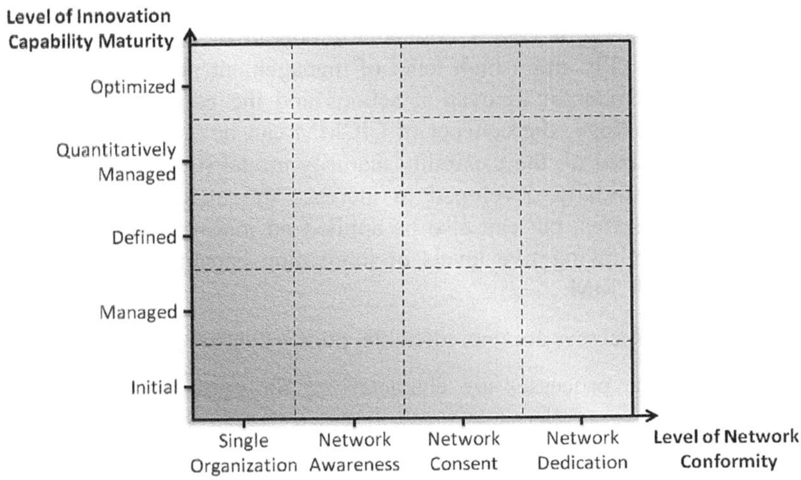

Figure 6.4. *Collaborative innovation capability maturity model [KNO 13]*

An initial proof-of-concept has been developed within the BIVEE project to account for the dimensions of change management, communication management, human resources and technology management. This methodology and its dimensions can be applied to assess the management maturity and conformance within a collaboration concerning collaborative innovation projects.

6.5. Conclusion and outlook

This chapter has described the BIVEE framework, consisting of the VEMF, BIRF and MF. Additionally, the CICMM has been introduced as a methodology to support the governing of collaborative innovation.

Although BIVEE defines the innovation space and production space separately, it has to be stated that both spaces are closely connected.

Innovations that are the outcome of the innovation space are transferred into the production space and experience made within the production space may be injected into the innovation space.

The project has developed entities and tools, such as the mission control room (MCR), virtual innovation factory (VIF) and PIKR, which allow transferring the approach of the BIVEE framework into industrial practice. This has resulted in multiple alterations of the framework, but is still an ongoing process that will continue in the future.

6.6. Bibliography

[CMM 02] CMMI Product Team, Capability Maturity Model® Integration (CMMI®), Version 1.1, CMMI® for Systems Engineering, Software Engineering, Integrated Product and Process Development, and Supplier Sourcing (CMMI-SE/SW/IPPD/SS, v1.1), Carnegie-Mellon Software Engineering Institute, Pittsburgh, 2002.

[COO 02] COOPER R.G., *Top oder Flop in der Produktentwicklung: Erfolgsstrategien: von der Idee zum Launch*, Wiley-VCH, 2002.

[DIE 97] DIENER E., SUH E., "Measuring quality of life: economic, social, and subjective indicators", *Social Indicators Research*, vol. 40, nos. 1–2, pp. 189–216, 1997.

[EDU 09] EDUM-FOTWE F.T., PRICE A.D., "A social ontology for appraising sustainability of construction projects and developments", *International Journal of Project Management*, vol. 27, no. 4, pp. 313–322, 2009.

[ESS 10] ESSMANN H., PREEZ N., "An innovation capability maturity model – development and initial application", *International Journal of Human and Social Sciences*, vol. 5, no. 1, pp. 44–55, 2010.

[HOL 01] HOLLING C.S., "Understanding the complexity of economic, ecological, and social systems", *Ecosystems*, vol. 4, no. 5, pp. 390–405, 2001.

[KIR 11] KIRCHMER M., "Reference models to empower MPE", *High Performance Through Process Excellence*, Springer Berlin, Heidelberg, pp. 87–101, available at http://link.springer.com/book/10.1007/978-3-642-21165-2, 2011.

[KNO 13] KNOKE B., "A short paper on innovation capability maturity within collaborations", *NGEBIS Short Papers*, pp. 7–11, 2013.

[KOT 08] KOTELNIKOV V., Systemic innovation, available at http://www.1000advices.com/guru/innovation_systemic_vk.html, [Accessed on 23 March 2012], 2008.

[STE 01] STEPHENS S., *Supply Chain Council & Supply Chain Operations Reference (SCOR) Model Overview*, Supply Chain Council, Inc., Pittsburgh, USA, 2001.

[THO 07] THOBEN K.-D., "LABORANOVA: a collaborative innovation vision (session 1: mechanisms for collaborative innovation in living labs)", *ESoCE Net Industrial Forum*, Rome, Italy, 3 December 2007.

[VAL 12] VALIENTEA J.M.A., AYERBEA G., FIGUERAS M.S., "Social responsibility practices and evaluation of corporate social performance", *Journal of Cleaner Production*, vol. 35, pp. 25–38, 2012.

[VON 06] VON GEIBLER J., LIEDTKE C., WALLBAUM H. *et al.*, "Accounting for the social dimension of sustainability: experiences from the biotechnology industry", *Business Strategy and the Environment*, vol. 15, no. 5, pp. 334–346, 2006.

[WAL 00] WALZ R., "Development of environmental indicator systems: experiences from Germany", *Environmental Management*, vol. 25, no. 6, pp. 613–623, 2000.

7

The BIVEE Environment: Description of the Overall Software Architecture

The BIVEE acronym stands for business innovation and virtual enterprise (VE) environment. The main tangible outcome of the BIVEE project is the ICT implementation of such an environment: a cloud-oriented software architecture, a cloud-ready platform providing a base layer of services, an integrated collection of end user applications and tools which enable business innovation and support value production. This chapter is an introduction to the BIVEE environment's architecture and a primer of the BIVEE platform from a technical perspective.

7.1. Introduction

The BIVEE project's spotlight is on two areas: *production space* and *innovation space*. Production space is where all activities related to the core business take place, and where business models and processes are designed, deployed, monitored and adjusted over time. At the same time, to counter the natural tendency of models and processes to become obsolete in a rapidly changing scenario, continuous open innovation is promoted and fostered in its own separate – but tightly connected – space. There is a continuous, bidirectional information flow between these two spaces: production feeds innovation with new challenges and innovation feeds production with new ideas.

Chapter written by Mauro ISAJA.

In the BIVEE environment's architecture, each space is covered by a dedicated web application targeted at end users: the mission control room (*MCR*) tackles production, while innovation is addressed by the virtual innovation factory (*VIF*). However, these rely on a common shared platform providing both with a base layer of data, knowledge, services and capabilities. The fully integrated solution – the two *BIVEE applications* plus the *BIVEE platform* – is collectively known as the *BIVEE Environment*.

The FInES[1] Cluster Research Roadmap has the following vision of a future Internet-based enterprise system: "*By 2020, the Internet will become a universal business environment on which new values can be created by competing as well as collaborating enterprises through innovation in a level playing field. Within this environment, FInES will become a transparent and invisible part of the business operation, available and affordable to all enterprises as required. It will make optimal use of the capabilities provided by a universal service infrastructure based on the concept of the Interoperability Service Utility (ISU)*" [FEC 10].

In the FI PPP[2] [EUC 11] context, and more specifically in the architectural blueprint of its core platform[3], a similar notion exists: the generic enabler (GE), which is defined as "*reusable and commonly shared functions serving a multiplicity of usage areas across various sectors*" [FWC 13].

While the concepts of ISU and GE differ slightly, they are both seen as the ultimate result of a commoditization trend in ICT. The universal service infrastructure envisioned in FInES, as well as the core platform in FI PPP, are basically the aggregation of several generic, stable and well understood building blocks – implemented by commodity software – into a feature-rich, integrated, universally-available environment where the new generation of future Internet applications can be developed and deployed.

The BIVEE environment can be considered as a small-scale experimentation of the Future Internet vision: specific goals are targeted by *vertical* BIVEE applications, which run on top of the *horizontal* BIVEE platform. Actually, the BIVEE platform aims at being the first prototype of a commoditized, future Internet-ready platform for an enterprise ecosystem.

1 Future Internet enterprise systems.
2 Future Internet public–private partnership.
3 The FIWARE project.

7.2. BIVEE environment reference architecture

The block diagram in Figure 7.1 illustrates, from a logical perspective, the reference architecture of the BIVEE environment, and the relationship existing between the BIVEE platform and the two BIVEE applications [DIB 12]. It also shows the composite structure of the platform: two layers of foundation – data persistence (i.e. raw data, meta-data and knowledge) and services – plus platform-specific capabilities such as knowledge and raw data management (PIKR and RDM, respectively), and a unified web UI (web portal). RDM tools, while an integral part of the BIVEE platform, are depicted as a separate element due to their substantially different technology[4].

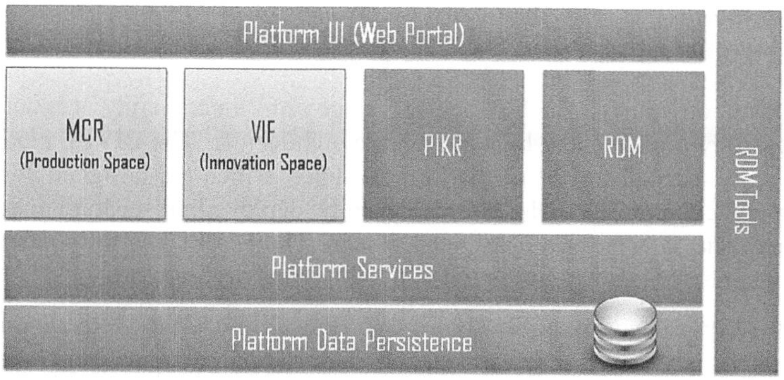

Figure 7.1. *BIVEE environment's reference architecture: the logical layers*

Given the main focus of this chapter, the two BIVEE applications – MCR and VIF – will only be briefly introduced, describing how they leverage the BIVEE platform. In the perspective of a wider use of the platform, these applications represent a typical but not limiting use case.

7.3. The BIVEE platform

In a handful of words, the scope of the BIVEE platform can be defined as "enterprise ecosystem interoperability". To achieve this goal, the platform

4 Personal desktop application vs. cloud-based – more information is provided the raw data connection services and tools chapter.

needs to address some basic requirements; from a top-level point of view, these can be categorized as follows:

– *Sharing of knowledge*: common repository for ecosystem semantics.

– *Harmonization of data*: collection and semantic lifting of information from ecosystem processes.

– *System integration*: common services supporting application-level interoperability.

– *Unified access*: centralized user entry point to a distributed application environment.

– *Federated security*: distributed user management with a centralized user authentication service.

Quite obviously, such a wide range of functionality cannot be successfully implemented by a monolithic architecture. The BIVEE platform is actually a collaboration of platform modules, which are designed as self-contained components. Modules can even be deployed on separate nodes of the network, to achieve a distributed system. The BIVEE platform modules, which are described in some detail in the following sections, are:

Production and innovation knowledge repository (PIKR) – Requirements addressed: sharing of knowledge, harmonization of data and system integration.

Raw data management (RDM) – Requirements addressed: harmonization of data, system integration.

Web portal – Requirements addressed: system integration, unified access, federated security.

7.3.1. *Production and innovation knowledge repository*

The PIKR module is a multi-tier application implementing a knowledge base, and enabling knowledge sharing across the ecosystem. Knowledge in PIKR is materialized as a set of ontologies and semantic annotation of knowledge resources, i.e. documents and data records which can be physically stored anywhere.

Ontologies and semantic annotations are maintained by PIKR as RDF triples (assertions in the form: subject, predicate, object) [W3C 04] in its triple store [TSD 12], which is PIKR's implementation of the knowledge repository layer as defined by the reference architecture (Figure 7.1). On top of this, PIKR also provides semantic reasoning and query rewriting services, the latter as a semantic-enhanced access path to the KPI data (see below). All these, as well as direct access to the triple store, are exposed to BIVEE applications through a web API, corresponding to the platform services logical layer.

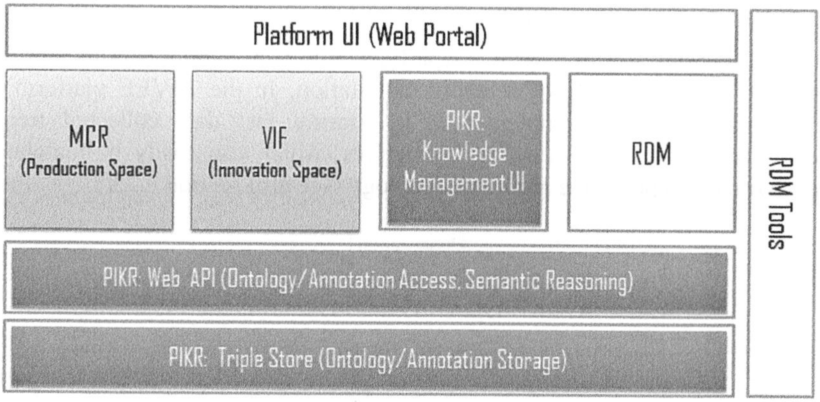

Figure 7.2. *PIKR components mapped to the BIVEE environment's logical layers*

As a practical example of PIKR's role in shaping ecosystem knowledge, we consider a specific BIVEE ontology which describes key performance indicators (KPI): *KPIOnto* [TSD 12]. KPIs are universally used to assess the performance of a process or of a whole business, with a strong focus on critical ("key") aspects like time and cost, customer satisfaction, employee happiness, environmental respect, etc. In BIVEE, KPIs are the final result of a complex data gathering and processing pipeline, where the mediation of KPIOnto is mandatory in order to collate raw data from heterogeneous sources. Raw data are defined as process indicator (PI) facts – i.e. data records representing atomic measurements.

As with other PIKR ontologies (here we can mention *DocOnto* for document-like resources, and *ProcOnto* targeted at business processes [TSD 12]), KPIOnto comes off-the-shelf as a basic set of classes, and is meant to be extended by the user to suit her own ecosystem needs.

Extensions are in the form of new subclasses and class instances. For example, KPI definitions are instances of the indicator ontology class, while KPI properties are subclasses of the dimension class. Once complete, KPIOnto represents a standardized, machine-readable model of the real world from which raw data originates: using this model, it is then possible to achieve a unified view of PI facts, regardless of their original source and structure.

For instance, a given KPI may be defined in KPIOnto as supporting *Organization Dimension* (Figure 7.3), which means that each single fact collected will have metadata attached describing how that value relates to standard entities like enterprise, department, project team, etc. This use of metadata is an example of semantic annotation. In the BIVEE platform's context, annotation is implemented by storing fact data collected from heterogeneous systems in *ad hoc* structures, which can easily be matched back with concepts of the reference ontology (see also section 7.2.2).

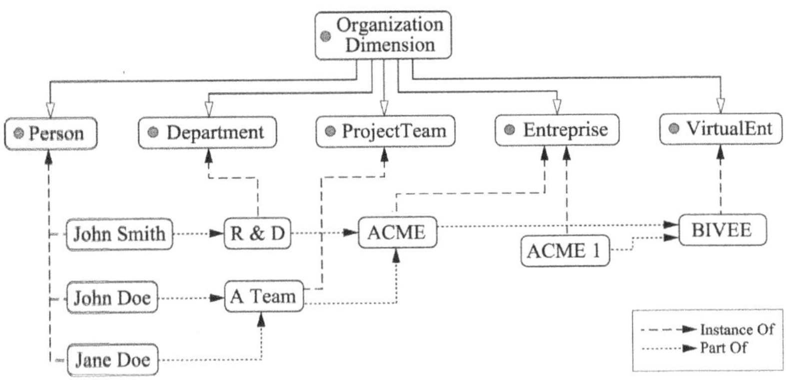

Figure 7.3. *Raw data management*

The RDM module addresses the need for collecting heterogeneous KPI raw data – PI facts – from ecosystem members, and making it available and understandable to consumer applications. The most relevant feature of RDM is its integration with the PIKR, which defines the KPIOnto ontology for the semantic enhancement of raw data (see also section 7.2.1).

In the RDM context, the data persistence logical layer of the reference architecture is implemented by the RDM-DS submodule (Figure 7.4). Facts are kept in *normalized* form – i.e. in a form that is highly compatible with the logical structure modeled by KPIOnto. The main value of data normalization is that application queries, which are expressed in terms defined by KPIOnto, can be automatically translated into equivalent RDM queries, which will perform very fast. By virtue of this mechanism, facts can be collected from heterogeneous sources – typically enterprise IT systems belonging to different organizations – and still be presented and analyzed in a homogeneous way. This ontology-driven approach to data interoperability is commonly called *semantic lifting*.

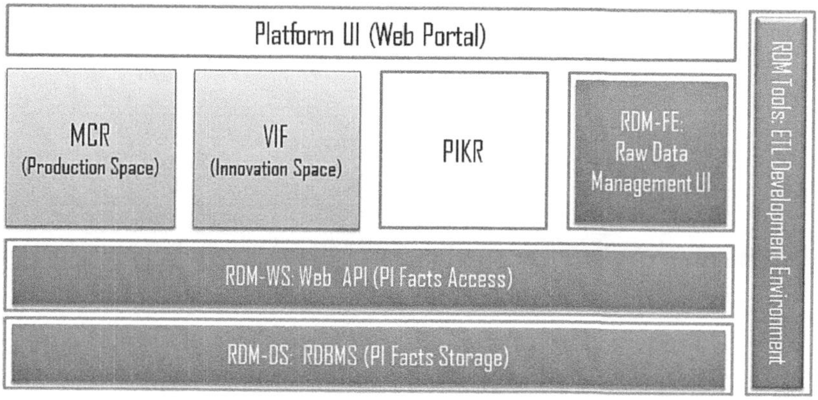

Figure 7.4. *RDM components mapped to the BIVEE environment's logical layers*

RDM-DS is implemented as an SQL database (MySQL Community Server[5]), with a specific data schema in which facts are stored. This schema is auto-generated, and continuously kept in-sync, using KPIOnto as a blueprint. Basically, KPIOnto classes derived from the dimension class are translated into domain tables following a set of rules (Figure 7.5), and are populated with records corresponding to concrete instances (Figure 7.6); each indicator instance becomes a fact table on its own, connected with the relevant domains. The resulting schema is described in the RDM-DS catalog, and from there entities can easily be mapped back to KPIOnto concepts [TSD 12].

PI facts are stored and published on the BIVEE platform by means of platform services, which in the RDM context are implemented by the RDM-WS submodule (Figure 7.4).

5 http://dev.mysql.com/downloads/mysql/5.6.html.

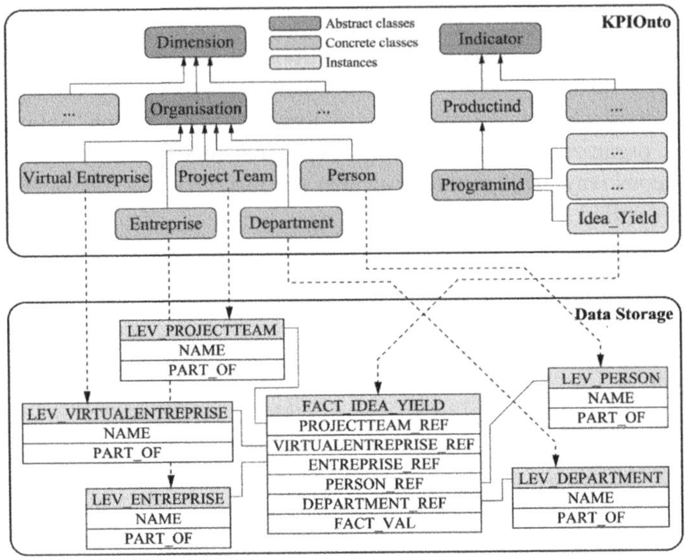

Figure 7.5. *Example of how KPIOnto concepts are mapped to RDBMS tables*

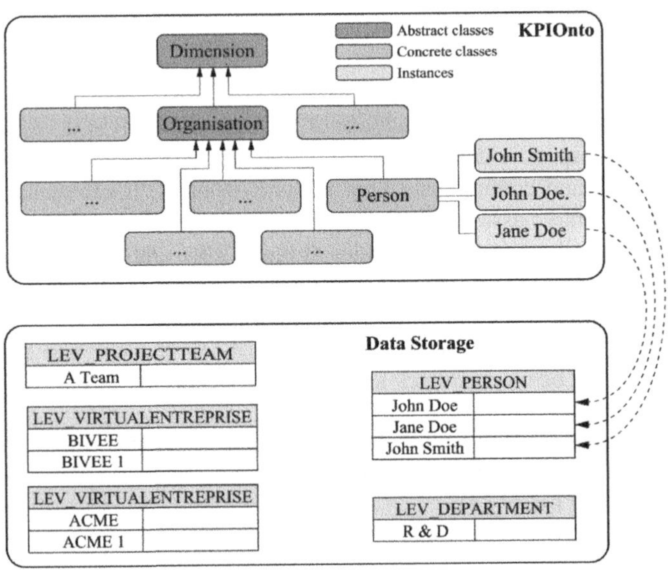

Figure 7.6. *Example of how KPIOnto instances are mapped to RDBMS records*

The collection of PI facts is implemented following a decentralized approach: each data contributor – typically an enterprise – designs, deploys and runs its own set of ETL batch procedures which extract data from ICT systems, transform them into a normalized form and finally load them on to the BIVEE platform by means of the RDM-WS platform services. This approach features some key advantages, the main advantage being that each data contributor runs the most computationally intensive operations on its own hardware, thus distributing the load across the entire ecosystem network.

To support this strategy, the BIVEE platform provides the raw data tools layer (Figure 7.4), implemented by the RDM-EDE (ETL development environment) submodule [TSD 13]. This is a desktop application for the assisted development of executable programs capable of feeding the platform services layer – i.e. RDM-WS – with normalized PI facts. RDM-EDE is meant to be installed on corporate workstations, as opposed to all the other platform modules which are typically run on the cloud.

RDM-EDE is an extension of the Talend Open Studio (TOS) for data integration open source IDE[6]. While this platform is targeted at the IT specialist, building ETL jobs does not require programming skills: it is mainly a matter of dragging predefined components on to a canvas, setting their properties and connecting them together to form a data processing pipeline. RDM-EDE extensions cover four aspects of development:

Meta-data synchronization – an integrated utility is used to keep RDM-EDE metadata in sync with the KPIOnto reference ontology. This allows users, at design time, to quickly and easily map their local data schema (the source) with the normalized structure of any given PI fact (the target).

Data translation – a dedicated functionality enables the user to set translation rules: how terms of the local vocabulary (the enterprise domain) are automatically transformed into their global equivalent (as per KPIOnto). As with schema mappings, vocabulary translations are defined at design time and applied at runtime.

Job composition – a custom component available in the job designed palette (tRDHOutput) implements the connection from the ETL job to the RDM-WS endpoint for PI fact loading.

6 http://www.talend.com/products/data-integration.

Job deployment – an integrated utility is used to deploy BIVEE-specific ETL jobs as autonomous executable programs (data translation dictionaries are bundled in the package).

7.3.2. *Web portal*

The web portal module provides a rich set of services and capabilities, focused on system integration. This is the most generic of platform modules, providing the lowest-level foundation elements on which the other modules and applications are built. With respect to the BIVEE environment's logical layers, it implements the following:

Portal registry [data persistence layer]: this is a platform-level registry where two kinds of entity persist: ecosystem users and ecosystem organizations.

Web API for application and module integration [services layer]: this set of REST services provides BIVEE applications and platform modules with direct access to the platform's user and organization registry. It also implements the Federated Single-Sign-On (see below) and the User Messaging service layers.

Web portal management UI [UI layer]: BIVEE administrators can use this web-based front-end to manage ecosystem users and organizations.

Centralized entry point/navigation of the BIVEE Environment [UI layer]: this is the actual web portal implementation, based on the open source Liferay Portal Community Edition software.

The federated single-sign-on (SSO) subsystem deserves special attention in this context, as it is a key element of the overall positioning of the BIVEE platform as an enabler for enterprise ecosystems [ISA 14]. Generally speaking, SSO provides a central authentication point for users of complex, distributed, multi-tier web applications and services: users authenticate themselves once and then are granted access to the entire environment, without repeating annoying login operations. There are two basic approaches at implementing SSO:

Centralized – all users, and their credentials, are registered in a single repository. This approach is most effective when used to centralize the management of security policies, e.g. within a single company. Each organization tends to keep an internal centralized repository of users and groups, often provided by a LDAP or Active Directory® server, in order to control access to corporate resources.

Federative – users and credentials are stored in local systems, or *nodes*, which are grouped in a federation. Accessing one of the federated nodes also grants access to all the others. This approach is particularly useful when different organizations want to keep control over their own security policies.

The federated SSO subsystem, as its name implies, follows the latter approach. It is designed on top of CAS[7] and SAML[8] open technologies, with some extensions:

1) the notion of VE and of enterprise as properties of the authenticated user, which implies the integration with the organization registry component of the portal data persistence submodule;

2) an attribute release feature: an SAML-based way of passing additional user information to SSO clients;

3) a standard data model for user attributes and roles.

This architecture is implemented by two platform services: the SSO service and the LSA service. As the diagram in Figure 7.7 suggests, all users are registered with a local security authority (LSA), which is an abstraction of a user repository exposed as a service. The LSA service is a two-faced object. On one side it exposes a public endpoint for user credential validation and user attribute release; on the other side, it connects with a pluggable user data repository. The SSO service, on the other hand, is centralized, and integrates all LSAs into a single system: user authentication and attribute release is performed here, but is actually delegated to the specific LSA which the authenticating user belongs to.

7 Jasig's Central Authentication Service, http://www.jasig.org/cas.
8 OASIS/SSTC Security Assertion Markup Language, http://saml.xml.org/.

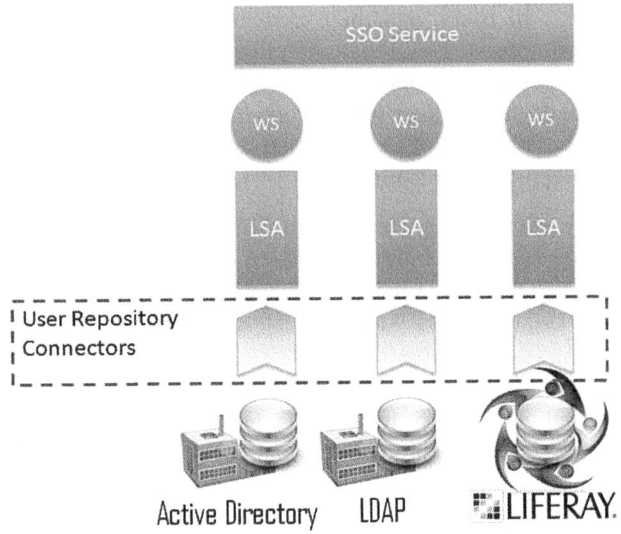

Figure 7.7. *Multiple LSA services under a single SSO master service*

7.4. BIVEE application: the mission control room

The MCR implements the production space logical layer of the BIVEE reference architecture (Figure 7.1). Its mission is to support stakeholders in the creation and management of VE. The following submodules are defined:

MCR-Modeler – VE creation, modeling and improvement.

MRC-Assistant – Collaborative environment for knowledge workers.

MCR-Monitoring – KPI-oriented dashboard for VE performance analysis.

All submodules are leveraging BIVEE platform services, in particular those related to ecosystem data and knowledge, i.e. the PIKR and RDM Web API. MCR-Assistant is also a consumer of services which are directly provided by the VIF Application (see section 7.4).

For additional information on MCR, see Chapter 8 of this book, and related documentation released by the BIVEE Project [KRC 12].

7.5. BIVEE application: the virtual innovation factory

The VIF implements the innovation space logical layer of the BIVEE reference architecture (Figure 7.1). Using the base layer of data, knowledge, services and capabilities provided by the BIVEE platform, the VIF supports the innovation waves of creativity, feasibility, prototyping and engineering as defined by the BIVEE framework [RKE 12]. Moreover, it provides advanced tools (innovation projects, idea wizard, shared whiteboard, collaborative document editor, collaborative mind mapper) and a semantic-aware crawler (observatory) for monitoring relevant information on the Web.

For additional information on VIF, see Chapter 9 of this book, and related documentation released by the BIVEE Project [WEK 12].

7.6. Conclusions

The BIVEE environment has been actively tested in the scope of the BIVEE project's *monitoring campaign*, i.e. the validation and impact analysis task which is run by end users AIDIMA and Loccioni in their own production environment.

At the end of the BIVEE project (December 2014), a cloud-based instance of the complete BIVEE environment was deployed as a proof-of-concept demonstrator of BIVEE's results [ISA 14]. This instance provides a future Internet environment as well: the BIVEE platform as an enabling infrastructure for enterprise ecosystems, on which to build domain-specific extensions and/or applications. The BIVEE platform captures and implements the low-level, cross-domain requirements of such enterprise ecosystem environments and translates them into the raw data, knowledge and services layers of its reference architecture. We believe that such architecture is well suited for future Internet applications in the business innovation usage area, and we expect that it may evolve into – or at least significantly contribute to – a future Internet commodity infrastructure.

7.7. Bibliography

[DIB 12] DI BERNARDO R., BIVEE Environment Architecture and Service Platform Specifications, technical report deliverable #6.20 of the BIVEE Project, pp. 21–24, 2012.

[EUC 11] EUROPEAN COMMUNITY, FI PPP Web Site. Available at http://www.fi-ppp.eu/, last accessed 15/06/2013.

[FEC 10] FINES CLUSTER, FInES Research Roadmap Final Report. Available at http://www.fines-cluster.eu/jm/Documents/Download-document/2-FInES-Cluster-2010-Research-Roadmap.html, last accessed 15/06/2013, p. 22, 2010.

[FWC 13] FI-WARE PROJECT CONSORTIUM, Overall FI-WARE Vision. Available at http://forge.fi-ware.eu/plugins/mediawiki/wiki/fiware/index.php/Overall_FI-WARE_Vision, last accessed 15/06/2013.

[ISA 14] ISAJA M., BIVEE Environment Prototype v2.0, technical report deliverable #6.33 of the BIVEE Project, pp. 22–28, 2014.

[KRC 12] KNOKE B., ROSSI A., CALLE F. *et al.*, State of the Art and Virtual Innovation Factory Design, technical report deliverable #4.1 of the BIVEE Project, 2012.

[RKE 12] ROSSI A., KNOKE B., EFENDIOGLU N. *et al.*, Specification of Business Innovation Reference Framework, technical report deliverable #2.2 of the BIVEE Project, 2012.

[TSD 12] TAGLINO F., SMITH F., DIAMANTINI C. *et al.*, Production and Innovation Knowledge Repository, technical report deliverable #5.1 of the BIVEE Project, pp. 31–32 & 80–94, 2012.

[TSD 13] TAGLINO F., SMITH F., DIAMANTINI C. *et al.*, Advanced Interoperability Services, technical report deliverable #5.3 of the BIVEE Project, pp. 34–48, 2013.

[WEK 12] WOITSCH R., EFENDIOGLU N., KNOKE B. *et al.*, State of the Art and Mission Control Room Specifications, technical report deliverable #3.1 of the BIVEE Project, 2012.

[W3C 04] W3 CONSORTIUM, W3C Semantic Web Wiki, “RDF specification”, http://www.w3.org/RDF/, last accessed on 15/06/2013.

8

The Mission Control Room

8.1. Introduction

Virtual enterprises (VEs) provide an environment where flexible production and corresponding service delivery are cooperatively and efficiently carried out by involved stakeholders who are focused on utilizing their core competences.

Value production space (VPS) is an environment which abstracts complex distributed reality of a VE. It considers different layers (e.g. operational, resource layer, etc.) and all the stakeholders involved (e.g. humans, services and machines), directly influencing the factors.

The mission control room (MCR) is the management suite, where the VPS is modeled with conceptual models considering aspects from the aforementioned layers. Moreover, MCR is the tool that offers functionalities by applying mechanisms and algorithms in models to create information values in order to set up a VE, monitor it and assist in its execution.

This chapter aims to introduce application scenarios, concept, realization and user experience of the MCR. Hence, this chapter starts by introducing application scenarios that are based on end-user requirements and are considered to be fulfilled by MCR. Secondly, it discusses the specified concept of MCR based on identified challenges, requirements and issues from application scenarios, and then the approaches and solutions in order to overcome these. Thirdly, it discusses how the MCR is realized and what technologies have been applied. Finally, in addition to a conclusion, user

---

Chapter written by Nesat EFENDIOGLU, Wilfrid UTZ and Robert WOITSCH.

experience made with MCR during the application of MCR in real settings is provided.

8.2. Application scenarios

There are three different application scenarios. For each scenario, different use cases have been considered, which should be supported by the MCR. The three cases are (1) virtual enterprise design, (2) virtual enterprise execution assistant and (3) virtual enterprise monitoring.

8.2.1. *Virtual enterprise design*

During the design of a VE, the MCR distinguishes between two application scenarios. The first one is a top-down scenario, which starts with the definition of the product (e.g. with its components, subcomponents and their attributes), for which the VE will be set up. The second one is a bottom-up scenario which consists of starting from concrete production processes and assuming a transformation from the existing processes.

However, MCR aims to support both scenarios, although the typical creation of VE may be the top-down approach and the typical improvement will be the bottom-up approach. MCR does not aim to limit the user, but wants to provide a flexible modeling support [WOI 12]. Hence, as shown in Figure 8.1, the following use cases are foreseen for an MCR modeler, which supports virtual enterprise design.

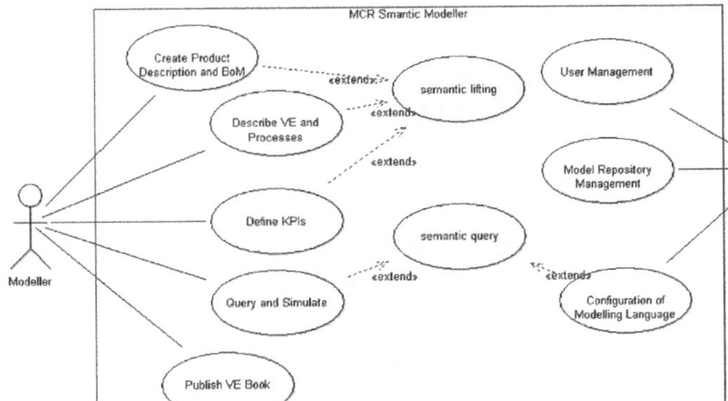

Figure 8.1. *Use cases for MCR modeler*

Furthermore, the semantic enrichment of an MCR modeler enables us to use a harmonized language while modeling and to annotate each object in a model by a predefined taxonomy of the Production and Innovation Knowledge Repository (PIKR) as introduced in [TAG 12]. Hence, every model designed with MCR will be semantically lifted with concept-defined ontologies in the PIKR.

Semantic queries are an extension of conventional queries that apply semantics to resolve the query, by semantic inferences to combine the results in an appropriate manner. For the end-user of an MCR, the query is transparent if it is handled by the MCR or passed onto the PIKR.

8.2.1.1. *Create your virtual enterprise by product definition*

As discussed in the top-down scenario, decision makers decide on the product and quantity, decompose the product into subproducts and define attributes such as quantity, cost and price. Hence, the MCR provides a product model by enabling us to specify material, components and a set of relevant features. Features can be freely chosen out of the predefined ones, according to its particular use.

Product specification leads to drafting the VE with respect to partner organizations by enabling semantic partner search, production and transport units, their service-level agreement, material, information and money flow, input/output and transformation tasks, as well as identification of relevant indicators by defining or selecting key performance indicators (KPIs).

Hence, the MCR provides a product model, a representation of the VE similar to a well-known e3 value model [GOR 15]. The VE process uses a notation like BPMN and a representation of KPIs.

8.2.1.2. *Improve your virtual enterprise*

Modelers design production processes on a more detailed level; using reference processes, they create use case processes. The production processes describe a sequence of processes performed within an enterprise; the so-called “swim lanes” enable the separation of processes performed by each individual partner within the VE.

Queries on the best fitting partners can be performed to query and simulate processes with alternative partner allocation. If necessary, additional changes are made in the description of the VE or KPIs.

8.2.2. *Virtual execution assistant*

As mentioned in [DAV 10], "knowledge workers are the key to innovation and growth in today's organization". Hence, knowledge workers are undoubtedly the driving force of successful "industrial" enterprises and are the key success factor in change management as they have the power to resist the competence managers and colleagues on which they rely on and, of course, their well-established ecosystem in which they have the power. MCR must therefore provide collaboration mechanisms for knowledge workers in order to participate not only during the VE creation but also during its continuous improvement. The process stepper engine enables the knowledge worker to step through the process via a Web platform.

As the user interface is designed to be used during execution phase, interactions with knowledge workers are realized by assistants. Hence, it consists of both a push-component and a pull-component. The push-component provides processes, a step-by-step explanation, documentation as well as contact persons, whereas the pull-component enables to feedback via Wiki, a quick poll or a mobile app. In that context, MCR distinguishes three different settings of process collaboration: (1) process training, (2) process testing and (3) process guidance.

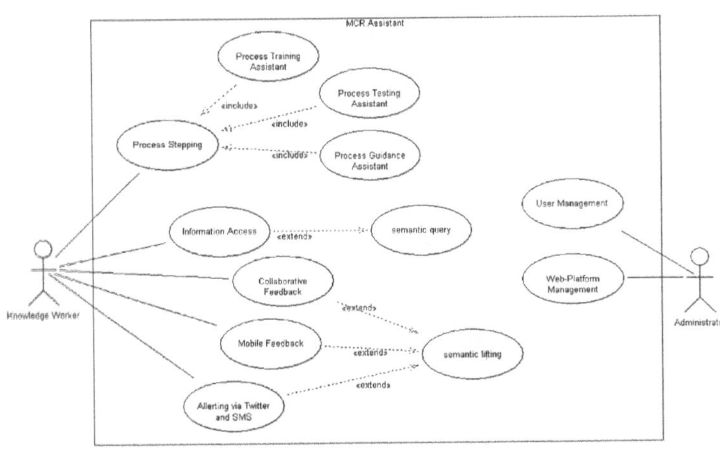

Figure 8.2. *Use cases for MCR assistant*

Figure 8.2 depicts the MCR assistant use cases, where the administrator deals with user management and Web platform management. The MCR assistant provides a process stepper enabling the knowledge worker to step through the process and receive information. MCR provides fixed relations between step, user and provided process information, and the semantic enrichment of the MCR assistant introduces loose coupling between these. Semantic query will be applied in a special configuration to enable loose coupling for information, users and process steps.

In case a knowledge worker provides feedback, it is a modeling task consisting of adding a "feedback object" into the model; hence, the same aforementioned mechanism of semantic modeling is applied to enable semantic lifting [HRG 13] of each feedback and alert.

8.2.2.1. *Process training*

In this setting, the knowledge workers can participate early in the process training, comment with their opinion and provide early feedback on the new processes. This setting is seen as early mock-ups of the VE process in which the knowledge workers can express their opinions.

8.2.2.2. *Process testing*

In this setting, the knowledge workers can comment and rate the processes. The goal is to assess the new processes by entering concrete KPIs in concrete situations.

8.2.2.3. *Process guidance*

In this setting, the knowledge worker is supported by the system in the form of a guiding and assistance system, and feedback is possible during concrete execution.

8.2.3. *Virtual enterprise monitoring*

MCR monitor is a hybrid presentation of predefined KPIs and a concrete set of values that are combined in an aggregated view to the user. Typically, this is presented to the decision maker.

The monitoring process consists of two parts. The first part is the presentation of the KPIs that are typically queried and viewed according to their context in the VE. The second part is the aggregation of concrete values and the possibility of drilling down those values. Both parts are presented in a cockpit that has two features: (1) the viewing of KPIs and (2) the selection and filtering of KPIs. Hence, the MCR monitor aims to support use cases as shown in Figure 8.3.

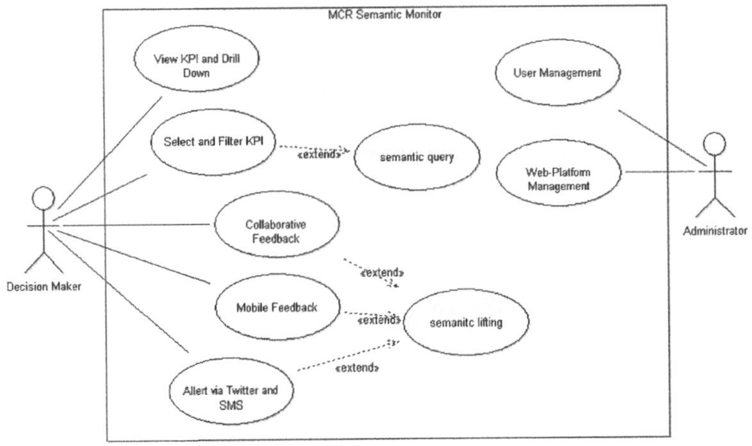

Figure 8.3. *Use cases for MCR monitoring*

The semantic enrichment of MCR monitor enables extended queries by using semantics and hence extending traditional KPI queries. This will be transparent to the user, as the user simply selects the filter, or search function, and is unaware of the fact that MCR corresponds to PIKR in order to improve the search.

8.2.3.1. *Define key performance indicators to observe production processes*

The modelers define business objectives, KPIs and production processes to be observed with KPI models provided by the MCR modeler.

KPI values are retrieved from PIKR with semantic queries containing annotations and production process contexts.

8.2.3.2. *Create your monitoring cockpit*

KPI viewing will be done by using the traffic light notation, where indicators are red, yellow or green. Besides this notation, there are alternatives, such as thermometers, balances, tachometers, charts or spider diagrams. MCR selected the traditional traffic light representation.

KPI selecting and filtering will be done in a tree-structure-like notation, where KPIs are structured according to a tree. Selecting a part of the tree results in the selection of all KPIs that are categorized accordingly.

Alternative selections are filters, search engines or configurations. MCR aims for a multiple-tree structure and a search engine that focuses on the relation of KPIs with other elements of the VE.

Similar to the aforementioned MCR modeler and MCR assistant, decision makers who analyze the MCR monitor are invited to provide feedback either for the modeler, using the same collaboration mechanisms such as the modeler, or for other decision makers using the alert feature.

8.2.4. *Collecting feedback through the whiteboard*

The MCR whiteboard is a special configuration of the MCR modeler and hence everything mentioned as semantic enrichment of the modeler holds true for the whiteboard.

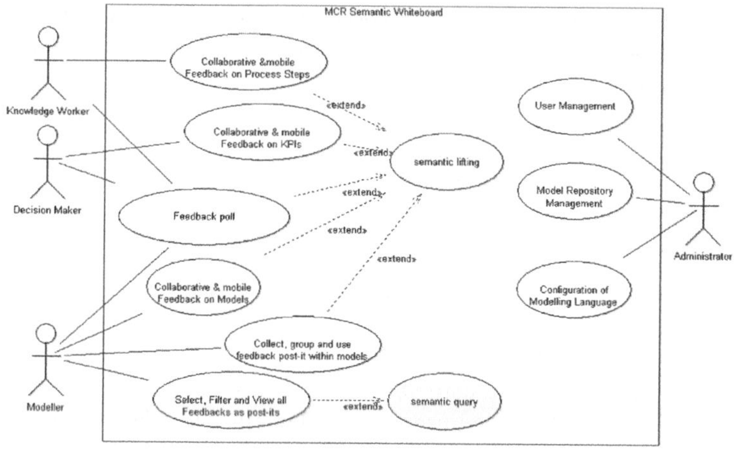

Figure 8.4. *Use cases for MCR whiteboard*

Figure 8.4 reflects the MCR semantic whiteboard use cases by highlighting that every feedback, poll or alert is semantically lifted and hence able to be semantically queried.

The semantic supported analysis of the feedback is probably one of the core advantages of the semantic enrichment of the MCR, as feedback, polls and collaboration input can be analyzed not only in the context of a concrete model, but also in the context of each other.

8.3. Concept

The VPS is seen as a real network of enterprises that agreed to form a VE. Hence, as described in section 8.2, MCR manages the VE by supporting the VE creation and improvement, collecting feedback of the knowledge workers during the VE execution, and providing features to monitor the VE and to enable creative processes for improvement or the injection of innovations.

MCR follows the concept model-based approach, which means that graphical concept models are used to chart the VPS and interpret these charts as a conceptualization of the real world. Hence, it is possible to apply a set of functions on that conceptualization, such as viewing, navigating, querying, simulating, publishing, training, testing, guidinges and collaboratively collecting feedback, assessing, alerting and improving.

In this context, in following section, we introduce identified requirements and issues that should be considered during the realization of MCR. Then, in the next section (section 8.3.2), we will discuss approaches and solutions to fulfill the requirements and solve the issues.

8.3.1. *Identified requirements and issues*

The key challenge is to enable such a conceptualization of the real world to enable (1) the semantic enrichment from the PIKR [TAG 12] and (2) innovation injections from the Virtual Innovation Factory [KNO 12].

We identify that VPS is a reality; hence, the conceptualization of this reality must consider (1) the state of the art in conceptualization, (2) currently applied approaches of the end-users, (3) a flexibility to adapt or change the conceptualization to the specific needs of the end-user and (4) the inclusion of unpredictable concepts using different formats, different world areas and different levels of semantic expressiveness.

MCR needs a conceptual framework, which is represented in the form of a modeling language, and a modeling tool, which is represented in the form of a Web application.

VPS is a highly complex socio-technical system including well-structured and production processes and knowledge workers. Hence, the conceptualization is required (1) to provide concepts that are sufficiently formalized enabling semantic support and (2) to strongly involve the knowledge worker and hence keep the "human in the loop".

8.3.2. *Approach and solution*

The MCR modeling method is specified by considering the following principles. The first principle consists of covering all the conceptual requirements. The second principle is aimed at introducing the plug-in concept enabling the currently identified set of modeling concepts in order to adapt the modeling language to alternative application scenarios and to stepwise extend if necessary. The third principle is to include semantic concepts to realize hybrid modeling. This means that concepts from the PIKR are imported into MCR in order to enable a user-friendly annotation of models, enabling the annotation in one tool.

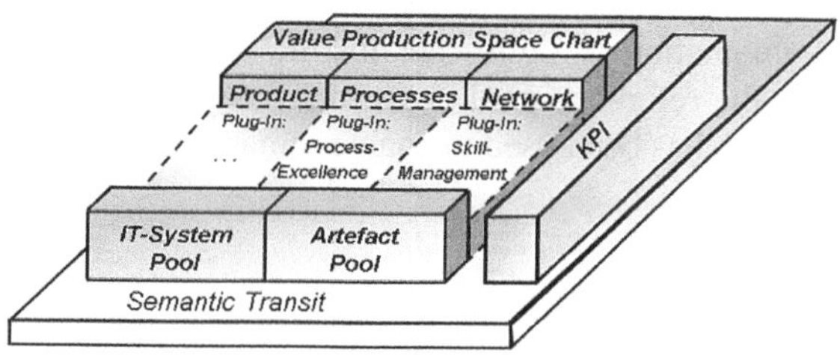

Figure 8.5. *MCR modeling method building blocks [WOI 12]*

Figure 8.5 introduces the building blocks of the MCR modeling method. As MCR provides the flexibility to create the VE using different paths, the core building blocks are provided to start either from product definition or from the already existing processes or network of the VE. Hence, these three building blocks of MCR modeling language are provided as either a detailed description or the starting point of the VPS.

A conceptual interface to more technical diagrams, such as workflows or software architecture – which are out of the scope of MCR – is provided by the building blocks the IT-system pool and Artifact pool. The IT-system pool identifies IT-system interactions with the business process. The Artifact pool is necessary to collect artifacts such as documents, which are required when using other MCR modeling building blocks.

Semantic transit building blocks allow for a conceptual interface with semantic systems like the PIKR. They also integrate as much semantics as necessary to have annotations and similar functionalities in one modeling tool.

Plug-ins enable the extension of the aforementioned building blocks by additional building blocks, concepts or functionalities.

Modeling stack, as an architectural design of MCR modeling language, has been introduced in [WOI 13]. Figure 8.6 depicts the so-called modeling stack of MCR modeling language [WOI 13]. The modeling stack defines the following building blocks: value production space chart representing geographical scope models as well as the "e3" value flow model; product representing the definition of products, their components as well as their subcomponents; process representing the processes in phases and levels comparable to supply chain operations reference (SCOR) model [API 15]; network representing the organizational structures and roles within the network of the participating enterprises in the distributed environment; KPI enabling the conceptualization of relevant parts of the concrete instances of the production processes; IT-system pool describing relevant IT-systems where user interaction or computation is significant for the designed processes; Artifact pool representing the reference pool to the relevant documents; and Semantic Transit Model, which has been used to enable semantic lifting [HRG 13].

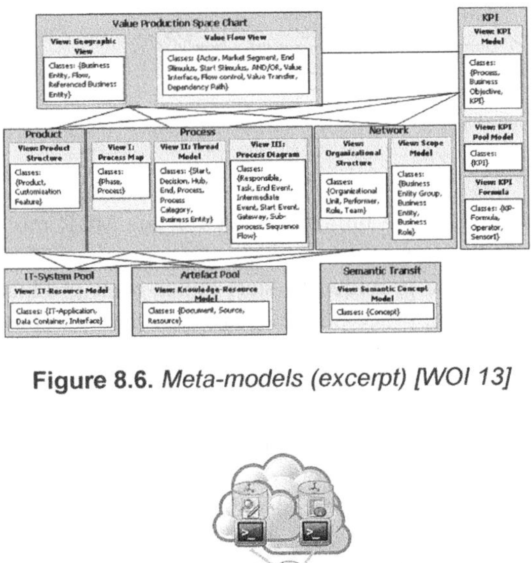

Figure 8.6. *Meta-models (excerpt) [WOI 13]*

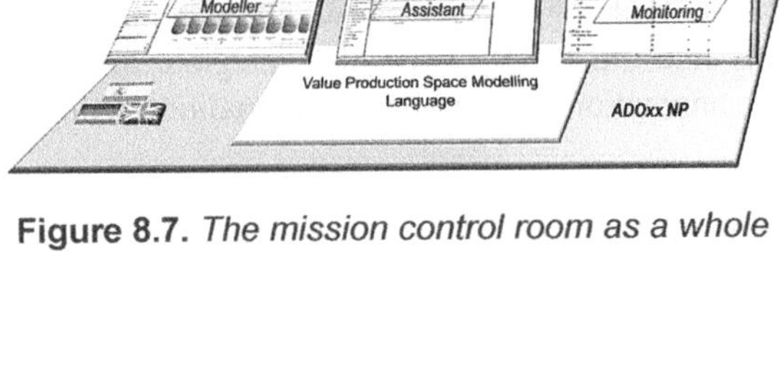

Figure 8.7. *The mission control room as a whole*

Figure 8.7 depicts the MCR as a whole, as a summary of the proposed solution for the aforementioned challenges, requirements and issues. Briefly, the solutions are as follows:

– domain-specific modeling language is developed that supports application scenarios, and the design, execution and monitoring of virtual enterprise;

– MCR is a modeling tool based on ADOxx® meta-modeling platform [ADO 15], which offers functionalities that can be used by MCR and also allows us to implement MCR-specific functionalities;

– foreseen components, MCR modeler, MCR assistant, MCR monitor, are developed to support use cases as they are introduced in application scenarios;

– MCR is a modeling tool based on Web technology, which provides full-fledged functionality over the Web;

– MCR enables collaboration, participation and networking, not through re-implementing required features but by using existing external tools via interfaces and integrating them with internal tool functionalities;

– realization of MCR is based on the ADOxxWeb framework, an ADOxx-specific plug-in concept, the so-called "method and functionality block" which provides a plug-in framework, corresponding deployment mechanism and monitoring mechanism to use conceptual and technical extensions;

– MCR provides Web services to be managed by other tools in case MCR functionalities are required to be integrated into other applications;

– MCR utilizes ADOxx® Identity Management and establishes interaction between CAS [JAS 15];

– MCR provides multi-language support.

8.4. Realization/technology

In order to realize approaches and solutions described in the previous section, a three-tier architecture (Figure 8.7) has been designed for MCR, whereas the ADOxx meta-modeling platform provides different support in all three tiers.

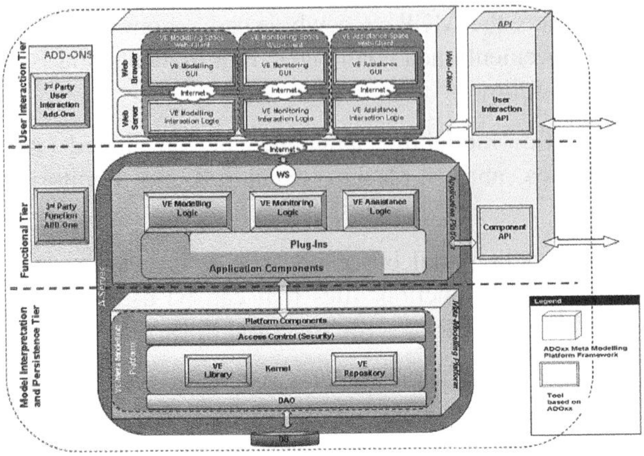

Figure 8.8. *MCR high-level architecture [DIB 12]*

In Figure 8.8, the overall MCR architecture is shown from the bottom to the top.

Data are stored in a database management sysetm (DBMS), whereas the database is twofold:

– based on the ADOxx meta2model, the meta-model database is structured and the concrete meta-model is stored as data in this database;

– based on the ADOxx meta2model, the instance database is structured and the concrete models, such as the reference models and the use case models, are stored as data in this database.

Hence, in order to access these instance data, a kernel that is aware of the meta-model – the so-called VE library – can interpret the instance data and thus can provide the VE repository.

In order to access the VE repository, the kernel needs to be accessed via a user interface layer and methods are provided in the form of a platform.

MCR depends entirely on the use of the plug-in concepts that are built on top of the platform. The functionality that already exists can be used either via inclusion of components or by inclusion of other plug-ins.

There are five plug-ins that constitute the functionality of the MCR:

– modeling plug-in provides all features that are typically required for a graphical modeling editor in order to view and edit models graphically;

– assistant plug-in provides features to train, test and propose processes to knowledge workers. Features that allow improving processes and collaborating with knowledge workers are provided;

– monitoring plug-in provides features to collect, aggregate and calculate KPIs. Hence, functionality to analyze, drill-down and search for KPIs, descriptions and concrete values are provided;

– Semantic Modeling Kernel (SMK) wraps relevant functional parts of the PIKR so that all other plug-ins in the MCR can use PIKR by accessing SMK. Besides the functional "wrapping", the PIKR must also "conceptually be wrapped";

– web service interface provides an application programming interface (API) to external tools that interact with MCR. It is introduced as a separate plug-in in order to provide external interfaces. External interfaces have the aim to simplify internal interfaces in order to ease the use of MCR form outside, without the necessity of insight knowledge.

These functional plug-ins contain business logic and follow the model control view (MCV) concept. The control part is taken over by a controller that is implemented in each plug-in to forward the method invocations. The models are populated mainly from platform interactions, as well as from Web service interface interaction and rarely from end-user interaction. The view is prepared and handed over to the user interaction tier.

The user interaction tier is applied in the Web environment by providing a Web framework. The Web framework creates the user sessions, and hence the user authorization to the user interaction tier, and establishes a framework for user interfaces. To achieve this, the framework relies on the single sign-on (SSO) and user identity management services made available by the business innovation and virtual enterprise environment (BIVEE) platform. More specifically, the Web framework integrates with the SSO service by means of a standard central authentication service (CAS) client, and delegates the management of user identities and credentials to the BIVEE application server, which is a Liferay instance.

Each user interface is implemented for each plug-in and activated by the Web framework. The whole Web framework can be wrapped into an iFrame and ported to a Liferay user interface. This is required in order to obtain a common user interface in the BIVEE platform.

The MCR consists of four main internal components: (1) MCR modeler, (2) MCR assistant, (3) MCR monitoring and (4) MCR whiteboard. The MCR whiteboard, which is actually a special view of MCR modeler, is equipped with some unique functionalities as introduced in application scenarios, and two main external components, Component API and SMK, as separated web services for third party use.

The creation of the internal components is based on the ADOxxWeb framework, as well as on additional web technologies. First of all, developing on the ADOxxWeb framework is referred to as developing plug-ins. Those plug-ins are referred to as method and functional building block (MFB) plug-ins which incorporate meta-model extensions (library), business logic, localization and look-and-feel, user interface structures, Web method declarations and Java code.

Given the architecture of the MCR, it is of great importance to understand the different programming application interfaces that are being used when developing ADOxxWeb plug-ins. Starting from a top-down approach, as depicted in Figure 8.8 with referring to used technologies, the following three main concepts are relevant:

– web interface: graphical user interface (GUI) development;

– web interaction: ADOxxWeb API for handling user-driven requests and providing basic ADOxxWeb customizing capabilities (altering the already offered functionalities);

– repository management: ADOxx API for offering full development support for new functionalities.

The GUI is built using the ExtJS framework [SEN 14], which provides full support for developing easy-to-maintain web graphical interfaces.

For defining the additional functionality, the ADOxxWeb API is utilized in order to customize the main functionalities and to provide an interface for communicating with the ADOxx server. ADOxxWeb API enables efficient layout and functionality customizing by allowing method and functionality overriding the modeling area.

The ADOxx API is used for establishing interactions with ADOxx Core functionalities such as repository management. This API enables full development capabilities by interacting directly with the MCR repository.

The MCR API development refers to the development of the Component API and the SMK API. The Component API provides a set of "get" and "set" methods that work with the MCR repository; however, this provided service resides outside of the ADOxxWeb environment and can be used by any allowed third party. The main functionality of the Component API consists of external link requests to ADOxxWeb which interprets the links using action handlers and returns to the requester the response to that request. To the same extent, the SMK interface is a specialization of the Component API, which offers MCR related methods (get all annotations of business objectives, get all goals with annotation, etc). The core implementation of the Component API and the SMK API relies on all the technologies mentioned in the previous paragraphs.

8.5. User experience

MCR has been applied in a VPS with real settings, and evaluated by technical and business experts. The user experiences about MCR have been reported in [SIN 14] and [AGU 14]. The following most significant comments and indicators are excerpts compiled from [SIN 14] and [AGU 14] that report the evaluation and validation results, which have been received from the users:

– MCR modeler: users comment that diagrams can be created quickly by this tool; however, some of the model types are very complex. They also think that the modeler is easy, clear and intuitive to use but some basics about modeling are required;

– MCR assistant: users comment that it is easy to understand how the MCR assistant works. Users consider that the different functionalities in the component are clearly defined and this seems to be an intuitive component. However, some problems in this component were found, such as unexpected warnings performing the step-by-step process or some issues related to the behavior of the zoom. These issues were reported and they have been addressed accordingly;

– MCR monitoring: users comment that it provides good overview on indicators, indicating possible problems and improvements. However, it is not easy to understand how the component works at the first glance.

Moreover, regarding the usability, the users have been asked to answer the following statements in Table 8.1 with "true" or "false". Answers indicate that MCR is attractive, intuitive to use and covers most of the needs of users with its functionalities; however, the performance of MCR has to be improved and some additional functionalities are required to satisfy users completely.

Sentences	Rate the following as True or False	User1	User2	User3	User4
S1	The starting screen of the application is attractive.	FALSE	FALSE	TRUE	FALSE
S2	The overall application is attractive.	FALSE	FALSE	TRUE	FALSE
S3	The application's graphics are pleasing.	TRUE	TRUE	TRUE	TRUE
S4	The site has a good balance of graphics versus text.	TRUE	TRUE	TRUE	TRUE
S5	The colors used throughout the application are attractive.	TRUE	FALSE	TRUE	TRUE
S6	The typography (lettering, headings, titles) is attractive.	TRUE	TRUE	TRUE	TRUE
S7	It is easy to find my way around the application.	TRUE	TRUE	TRUE	FALSE
S8	I can get to information quickly.	FALSE	TRUE	FALSE	FALSE
S9	It is easy to remember where to find things.	TRUE	TRUE	TRUE	TRUE
S10	Information is layered effectively on different screens.	TRUE	TRUE	TRUE	TRUE
S11	Information is easy to read.	TRUE	TRUE	TRUE	TRUE
S12	Screens have the right amount of information	TRUE	TRUE	TRUE	FALSE
S13	I always felt I knew what it was possible to do next	FALSE	FALSE	FALSE	FALSE
S14	It is clear how screen elements (e.g., pop-ups, scrolling lists, menus) work	FALSE	TRUE	TRUE	TRUE
S15	My mistakes were easy to correct.	FALSE	TRUE	FALSE	FALSE

Table 8.1. *Results of Ratings about Usability of MCR [AGU 14]*

8.6. Conclusion

In this chapter, we have presented a management suite – the so-called MCR – which can be used to model VPS with conceptual models considering aspects from different layers and which offers functionalities in order to set up a VE, monitor it and assist in its execution by creating information values from conceptual models.

We presented, respectively, application scenarios, identified challenges, requirements and issues, conceptual and technical specification of the MCR, applied technologies during the realization of MCR and, finally, the experience made by and the feedback provided by the users of MCR.

Although MCR is a research prototype, which is implemented as proof of concept, collected feedback from users leads us to future works, such as cooperative decision possibility, simplified modeling and widening semantic enrichment.

8.7. Bibliography

[ADO 15] ADOxx.org, available at www.adoxx.org/live/home, 2015.

[AGU 14] AGUIRRE M., EFENDIOGLU N., GIGANTE F. *et al.*, Evaluation assessment report, BIVEE EU-Project Deliverable 3.30, 2014.

[API 15] APICS SUPPLY CHAIN COUNCIL, SCOR, available at www.apics.org/sites/apics-supply-chain-council/frameworks/scor, 2015.

[DAV 10] DAVENPORT T., "Process management for knowledge work", in VOM BROCKE J., ROSEMANN M. (eds), *Handbook on Business Process Management 1, International Handbooks on Information Systems*, Springer, 2010.

[DIB 12] DI BERNARDO R., SMITH F., DIAMANTINI C. *et al.*, BIVEE environment architecture and services platform specifications, BIVEE EU-Project Deliverable D6.20, 2012.

[GOR 15] GORDIJN J., AKKERMANS H., E3 value model editor, available at e3value.few.vu.nl, 2015.

[HRG 13] HRGOVCIC V., KARAGIANNIS D., WOITSCH R., "Conceptual modeling of the organisational aspects for distributed applications: the semantic lifting approach", *IEEE Annual Computer Software and Applications Conference Workshop (COMPSACW)*, 2013.

[JAS 15] JASIG, CAS, available at http://jasig.github.io/cas/4.0.x/index.html, 2015.

[KNO 12] KNOKE B., ROSSI A., CALLE F. *et al.*, State of the art & virtual innovation factory design, available at bivee.eu/download, 2012.

[SEN 14] SENCHA, Ext JS 4.2.2 Sencha Docs, available at http://docs.sencha.com/extjs/4.2.2/, 12 February 2014.

[SIN 14] SINACI A.A., PACACI A., ROSSETTI G. *et al.*, Results of the monitoring campaign (r2.0), BIVEE EU-Project Deliverable D7.31, 2014.

[TAG 12] TAGLINO F., SMITH F., PROIETTI M. *et al.*, Production and innovation knowledge repository, BIVEE EU-Project Deliverable D5.1, 2012.

[WOI 12] WOITSCH R., EFENDIOGLU N., BENJAMIN K. *et al.*, State of the art and mission control room specification, Deliverable d3.1, BIVEE European project, 2012.

[WOI 13] WOITSCH R., SMITH F., TAGLINO F. *et al.*, "Semantics-based enhancement of value production conceptual models: a pragmatic approach", *eChallenges 2013 Proceedings*, 2013.

9

The Virtual Innovation Factory

9.1. Introduction

The virtual innovation factory – VIF for short – is a factory that does not operate on solid goods, but on knowledge with the purpose to create and elaborate ideas toward innovations. The VIF concept is realized with the design of platform that is based on the scientific foundation of the BIVEE framework and an analysis of existing innovation management platforms.

It is a special kind of factory that helps generating ideas and aims at managing innovation, not just inside one company or organization, but in many of them. In the context of BIVEE, we call these agglomerations of actors, virtual enterprises. In a way, we can say the VIF produces and manages innovation by motivating and encouraging users (employees, students, professors and researchers) of these enterprises/organizations in participating (and taking an active interest) in the innovation cycle of their own organization.

The VIF aims at generating innovative ideas that will produce value, once managed and validated on a business and operational level, in the production space. The VIF is, therefore, based on knowledge and creates innovation paths or plans from which the ideas and innovation itself are managed.

9.2. Methodological background

During the early stages of the BIVEE project, a reference framework was created. In many areas and aspects, this BIVEE framework does not want to

---

Chapter written by Francisco CALLE MORENO.

reinvent the wheel on a theoretical level, but uses and adapts well-known scientific theories and models (e.g. SCOR, VRM and waterfall model). It is innovating in the aspects that are relevant to the project, which are mostly related with the virtual enterprise.

For the innovation space, the related reference framework is the business innovation reference framework (BIRF). The BIRF aims to provide an overall theoretical structure and an overview of the main pillars and components that describe the actions within the VIF. Instead of following a more traditional approach like linear workflows, in BIVEE there have been defined four phases of activities named *waves*. Each wave contains *a certain number of activities* to be carried out, but this is not always true. Sometimes the order is not relevant for a wave to be completed. Each activity also contains *a set of documents* to be filled in order for each activity to be completed. Each document is associated with certain *key performance indicators* (KPIs) to be first set by the virtual enterprise and then met in order for the document to be considered complete in a satisfactory way for the given idea that is progressively shaping into a project. In order to give some guidance to the innovation space, the BIRF is organized into four main waves:

– *Creativity wave*: this wave starts with an innovation idea or a problem to be solved, providing an initially sketchy idea to be developed and put into the business context of the virtual enterprise undertaking it.

– *Feasibility wave*: this second wave provides and clarifies with elements and information in order to justify in economical and operational terms the actual undertaking and the further development of the original idea.

– *Prototyping wave*: this wave puts into concrete actions the previous collected information and elements producing a first implementation of the initial idea in the form of a prototype.

– *Engineering wave*: in this wave, the original idea transformed into a prototype is attentively analyzed to generate production and engineering plans.

9.3. Current status

9.3.1. *Baseline*

The VIF basis was a Drupal-based platform designed and developed by Atos, named PGI, which is a Spanish acronym for *Plataforma de Gestión de Ideas,* or idea management platform. During the first year of the project, the development team decided to drop the current baseline. The reasoning for this cumbersome decision was:

– the current platform was not flexible enough to be adapted to the changes happening in the ICT world;

– its technology offered no improvement against current trends and functionalities;

– it lacked backward compatibility, as it could not work with input generated by an older product or technology. Previous versions of the same platform produced several more problems than benefits when updating to newer versions;

– as an ICT project, the team felt that there was a need to research some of the new emerging technologies and apply them to the project. In this particular case, web-based applications in real time;

– to apply technologies, which are as homogenous as possible. Homogeneity in technologies leads to lesser resource requirements for their development. This is especially important for SMEs;

– last but not least, the restless minds of the development team that are in a continuous search for the next big hit, in this case, the next best framework.

9.3.2. *Technology change*

The trend in web applications has changed a lot in the last few years: from Java to Python and Ruby. However, PHP has always been the favorite option for web developers. In recent years, an old but renewed technology has arisen as an alternative to all of them: JavaScript.

The reasons for choosing JavaScript are:

– it is freely available. JavaScript is arguably the most open programming language. There is ECMA-262 [ECM 11], its specification is an ISO standard. This specification is closely followed by many implementations from independent parties. Some of those implementations are open source. Furthermore, the evolution of the language is handled by TC39 [ECM 12], a committee comprising several companies, including all major browser vendors;

– JavaScript can be used for the user interface and, nowadays, also for backend processes due to technologies such as Node.js;

– in the area of graphical user interfaces, JavaScript benefits from being part of HTML5 [W3C 14]. HTML5 is deployed widely and making constant progress. It is slowly becoming a complete layer for writing full-featured, cross-platform applications. Thereby, the Java platform is almost like an embedded operating system. One of the HTML5's selling points is that it lets you write cross-platform graphical user interfaces. In the past, "cross-platform" meant Windows, Mac OS or Linux. But now, there are two additional interactive platforms: web and mobile;

– apart from the user interface, JavaScript is used for:

- Libraries. JavaScript has an abundance of libraries, which enable us to complete tasks. From processing and displaying PDF files (via PDF.js) to displaying vector graphics,

- Backend. The Node.js platform allows developers to write server-side code and shell scripts (build tools, test runners, etc.),

- JavaScript Object Notation (JSON) [ECM 13]. JSON is a data format rooted in JavaScript that has become popular for exchanging data on the Web, i.e. the results of web services,

- NoSQL databases (such as CouchDB and MongoDB) [DBE 13]. These databases tightly integrate JSON and JavaScript.

– JavaScript has lots of tools and is getting better building tools (e.g. Gulp) and testing tools (e.g. mocha). Node.js makes it possible to run these kinds of tools via a shell (and not only in the browser). One risk in this area is fragmentation, as there are progressively too many of these tools developed;

– JavaScript is fast. JavaScript engines have made tremendous progress, evolving from slow interpreters to fast just-in-time compilers. They are now fast enough for most applications. Furthermore, new ideas are already in development to make JavaScript fast enough for the remaining applications.

Given those arguments, JavaScript is doing remarkably well. It is certainly not perfect but, at the moment, it is hard to beat – and things are only getting better.

9.3.3. *The selected framework*

At the moment of the change, a SoTA was made in order to select the best possible framework. The final candidates were AngularJS, Ember.js and Meteor. After several tests with all of them, lots of internal discussions and analysis of a large amount of posts[1] in forums discussing the same topic, a decision

1 A list of JavaScript frameworks can be found at http://www.todomvc.com.

was taken: Meteor had to be the selected framework. Meteor is a client-database-server framework (full stack), written entirely in JavaScript (homogeneous). It takes care of almost all of the plumbing for the developer, allowing them to concentrate on developing business value:

– from an end-user standpoint, Meteor-based apps bring a desktop-like feeling with no reload and instant synchronization among connected users. Also, latest releases offer integration with portable devices such as smartphones. This is the base for future versions of the platform that can be used everywhere;

– Meteor is an open-source, real time and cross-platform web application framework built on top of proven technology that allows for very rapid prototyping and produces cross-platform (web, Android and iOS) code. It is also production-ready;

– it offers a lot out of the box, while working with other technologies in the ecosystem;

– the Meteor Development Group has received $11.2M in funding from the venture capital firm Andressen Horowitz, which has also invested in Twitter, Airbnb and Foursquare [DAR 12];

– investing in Meteor is indisputably worthwhile, given the framework's developer base and momentum. Version v1.0 was launched on 28 October. November 1st, Meteor surpassed 21,000 stars on github, becoming the 11th most starred project on GitHub, with over 4,000 developers from 130+ cities in 40 countries gathered on 6 November 2014 to celebrate Meteor day – a global event where BIVEE was present;

– by contrast, the next most popular full-stack JavaScript framework, Derby.js, has 1/10th the footprint on StackOverflow, GitHub and Quora. AngularJS (albeit for the front-end only) is being radically revamped in the upcoming 2.0 version, and "users will need to get to grips with a new kind of architecture. It has also been confirmed that there will be no migration path from Angular 1.X to 2.0" – this was a general concern.

9.3.4. *A more technical overview of the selected framework: Meteor*

Meteor works with existing front-end libraries (i.e. React, Angular, Famo.us), and its own front-end library – Blaze – is perfectly integrated into the framework and fulfills the same purpose as Angular, Backbone, Ember,

React, Polymer or Knockout, but is much easier to use (and can be used with any of these libraries if you have to). Meteor works with existing back-end via REST APIs if needed, but is better when using DDP [GIT 14] (a client-server protocol for querying and updating a server-side database and for synchronizing such updates among clients. It uses the publish-subscribe messaging pattern), a simple and powerful publish/subscribe protocol. The server side runs on top of Node.js and eliminates call-back problems by using fibers [RAM 12]. The underlying database is MongoDB. Meteor includes real time and highly efficient monitoring for changes to data via oplog tailing. It monitors MongoDB's oplog to avoid extra operations on the database. This significantly reduces the number of queries needed to obtain the freshest changes to the MongoDB.

Meteor has two types of servers, as shown in Figure 9.1. One for HTTP and another for DDP:

– the HTTP server is used to serve static files and HTTP requests. Meteor uses the connect Node.js module for this purpose;

– the DDP server handles all publications, MongoDB operations and Meteor methods. Meteor uses SockJS as the transport protocol. The DDP server is a SockJS server customized for Meteor.

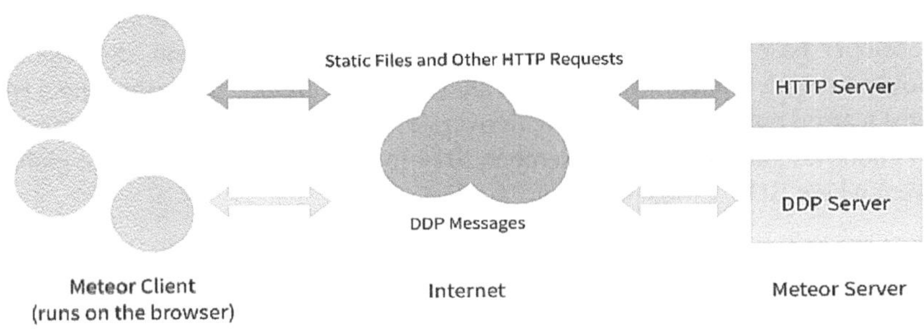

Figure 9.1. *Meteor framework layout*

Meteor is built on top of the MongoDB – which is not a real-time database – but Meteor is real time. Meteor makes MongoDB real time using two techniques:

– polling MongoDB every 10 s;

– using MongoDB oplog [MET 14].

Polling is a very expensive operation and this is why Meteor includes another option (using oplog). It needs some additional setup which is not possible with shared MongoDB hosting services and, therefore, it is not used for the prototypes since the amount of data handled does not require this, but is a starting point for possible future developments.

The integration of Meteor and the architecture of the VIF are shown in Figure 9.2. To summarize, there is one central access point (1) which is the main UI. The database (2) can be stored in a cloud-based provider as it is now and offers, if needed, direct access to the data through REST APIs (3). The VIF can use third-party APIs (4) such as Disqus for the comment system but using our data stores. Finally, VIF components (5) are designed to run as standalone applications and also integrated with the main UI (1) using REST-based services or shared database access. The latter is the approach followed in this project.

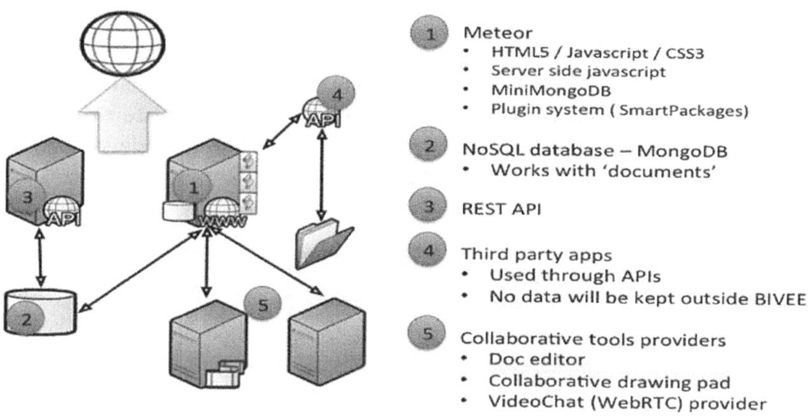

Figure 9.2. *VIF architecture overview*

Finally, and to better understand where those technologies were applied, Figure 9.3 explains the technology-based VIF structure. The technology stack is based on current standards or W3C recommendations and has reduced its complexity during the project. The list of technologies used can be reduced to HTML, Javascript, CSS and MongoDB.

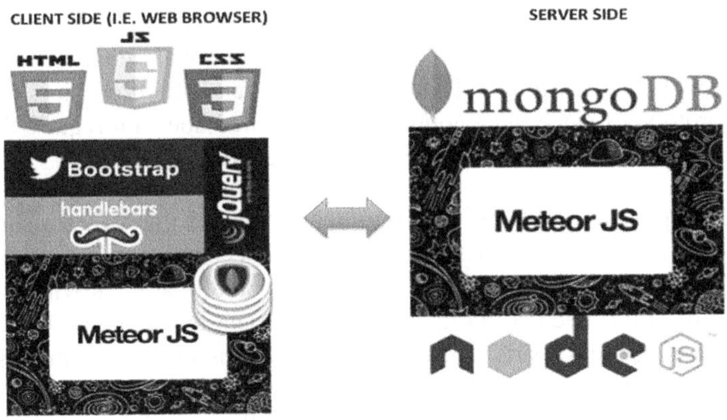

Figure 9.3. *VIF technologies*

9.3.5. *Components*

Based on the described technologies, the BIVEE project has developed the VIF. The VIF is built based upon some individual components. These components are shown in Figure 9.4. Some of them have been developed within the BIVEE project and some others have not yet been developed due to time restrictions, but are meant to be developed in the future.

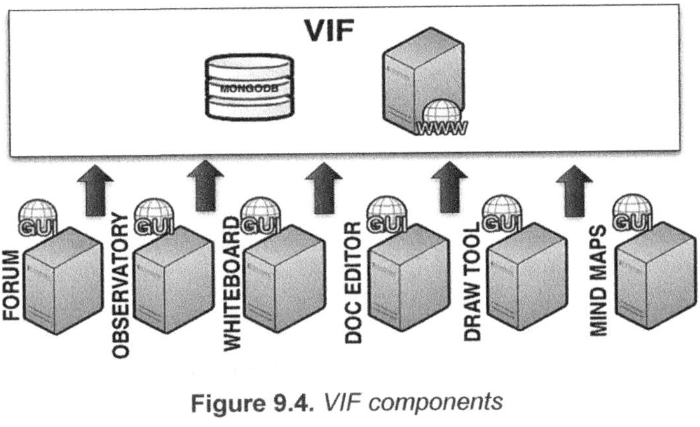

Figure 9.4. *VIF components*

The next sections describe in detail the key components of the VIF. Some of the components present in Figure 9.4 are merely concepts and yet to be finalized and polished before they become part of the future VIF.

9.3.6. *The main VIF application*

The VIF main application is the central point, where all the resources provided for the innovation process are accessed. Its main functionalities are:

- idea gathering and fostering;
- innovation project creation and follow-up;
- access to collaborative tools.

As described before, creativity starts with an idea that may address an issue. It can, of course, be an idea that is totally not related to an issue and be something new. At the time of idea creation, the owner of the idea – its creator – can invite other members of the platform to work together on it with the objective to create a better idea. Since *inspiration* can arrive at any time anywhere, the VIF is designed to work on any device using responsive design, which means that the application user interface will be adapted to the device that it is been displayed on. This includes not only desktop browsers, but also mobile devices such as smartphones and tablets. Once the idea is fulfilled, it needs to be annotated in order to provide knowledge to the central knowledge repository (the PIKR) and to find possible related contents or duplicates.

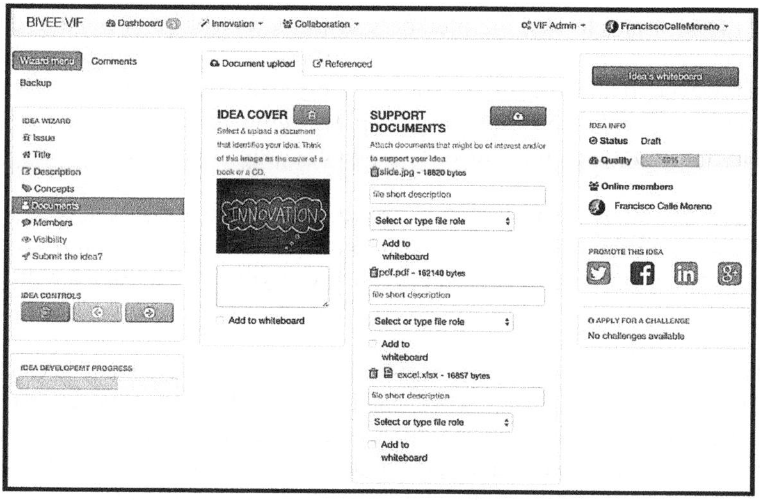

Figure 9.5. *VIF idea wizard, step documents*

The final step is to submit the idea to the domain experts, a role defined in the BIVEE platform. These members will evaluate the idea, comment on it and provide an answer whether the idea is accepted or rejected.

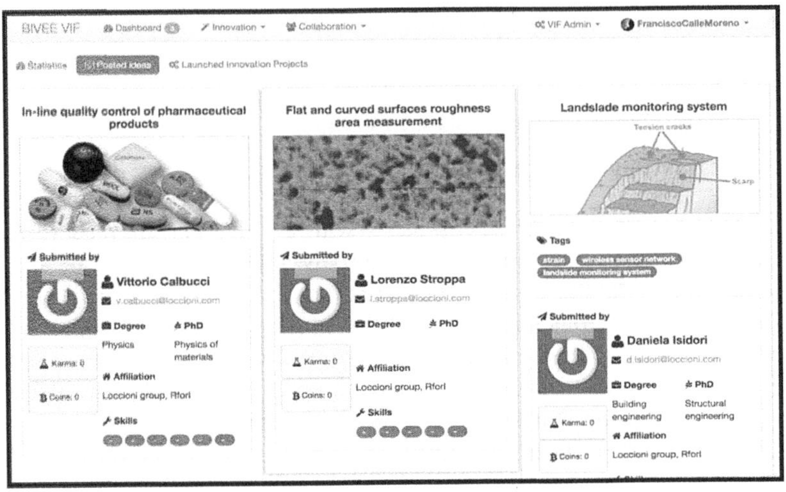

Figure 9.6. *Posted ideas to be evaluated as possible innovation projects*

If the idea is rejected, the members of the idea will be notified and are then required to decide if the idea will be refined to be submitted again, or not. However, if the idea is accepted, a new innovation project is created with, at least, the domain expert as the leader of the project and with the idea owner – collaborators on the idea can also be selected to be added.

Once the innovation project is created, more members can be added. They can be different to the ones contributing to the idea creation – to the project based on their skills. Finally, when everything is set up, the project can start.

The project is managed internally with a component designed specifically for BIVEE: the wave engine. It handles the lifecycle of the project open and closing the so-called gates – key points where a decision to transit to a different wave has to be taken. To support this decision, the VIF, in collaboration with other BIVEE components, records metrics (KPIs).

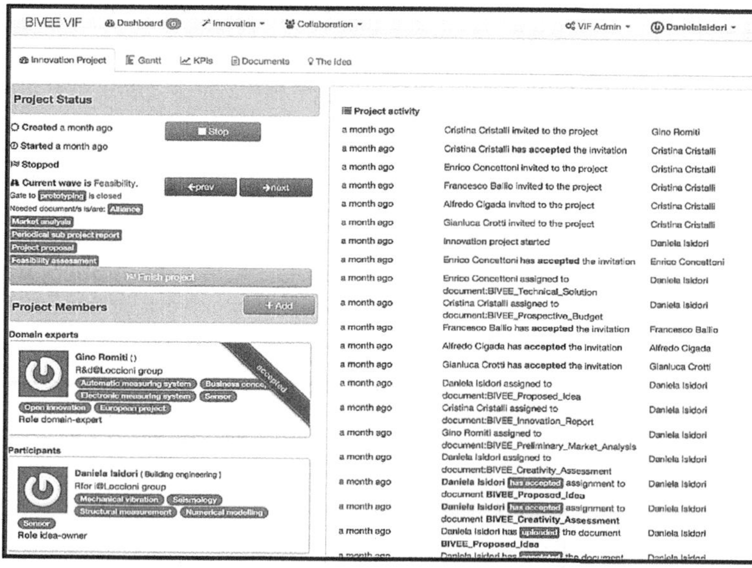

Figure 9.7. *Innovation project main screen*

When the project is finished, the full set of documents is exposed so the production space can apply the innovation.

9.3.7. *Fostering creativity*

One of the aspects of the BIVEE project is its human-centric approach. Therefore, the project is aware that creativity should be promoted and cannot be expected to flourish by itself.

In the pursuit of this goal, the VIF has been equipped with tools that can encourage its users to engage in active participation. Some of them were left on a theoretical level, and others were developed. The two proposed tools were the *VIF game & price* and the *Call4Ideas*.

The first tool was partially developed with the intention to provide the members of the platform with a game (*gamification*[2]), where they could

2 Gamification (or serious games) is the use of game thinking and game mechanics in non-game contexts to engage users in solving problems and increase users' self-contributions.

invest their BIVEE coins on ideas and innovation projects, in order to create a stock-trade-like game. Depending on the results of their investments, their enterprises in the platform would offer prices in many ways – physical prices, additional time, etc.

The Call4Ideas is a typical approach to collect ideas. The calls are always linked to a topic or an issue and can be created in two ways: as a single call, with the intention to participate in an innovation project and be a part of it and its possible outcomes; the other is a challenge with defined prices where users send their ideas and the selected ones will receive a price, but they cannot be a member of the innovation project.

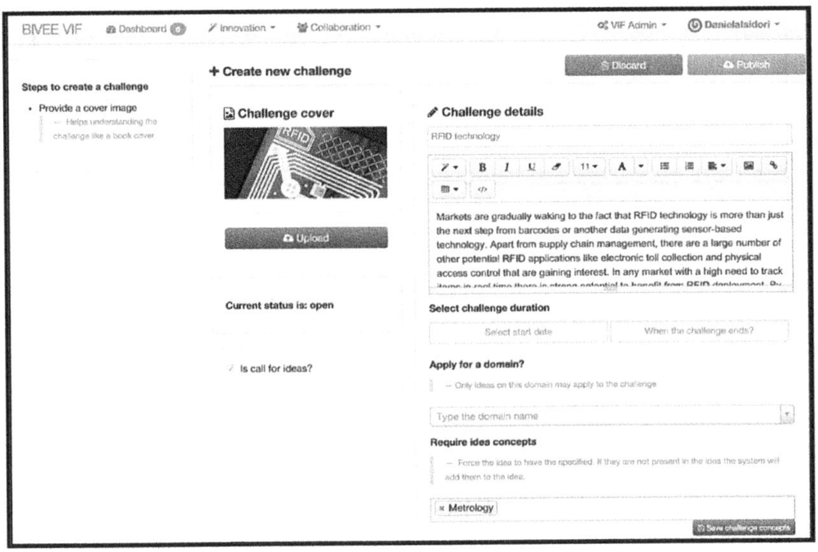

Figure 9.8. *Call for ideas creation*

9.3.8. *Collaborative tools*

Finally, since the VIF is a real-time web-based application, some collaborative tools are provided. Although some of these tools were declared as out of scope, two tools were provided: the meeting room and the collaborative document editor. Both have been integrated into a single tool.

The meeting room allows the members of the platform to have real-time video conferences and to collaborate in documents having declared an agenda in advance, open issues and to take private notes. Screenshots of the meeting room tool are provided in Figures 9.9–9.11.

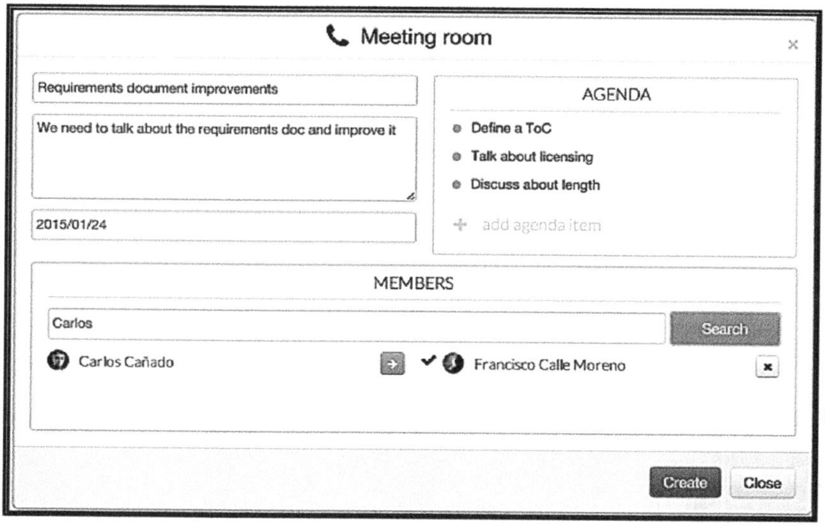

Figure 9.9. *Meeting creation*

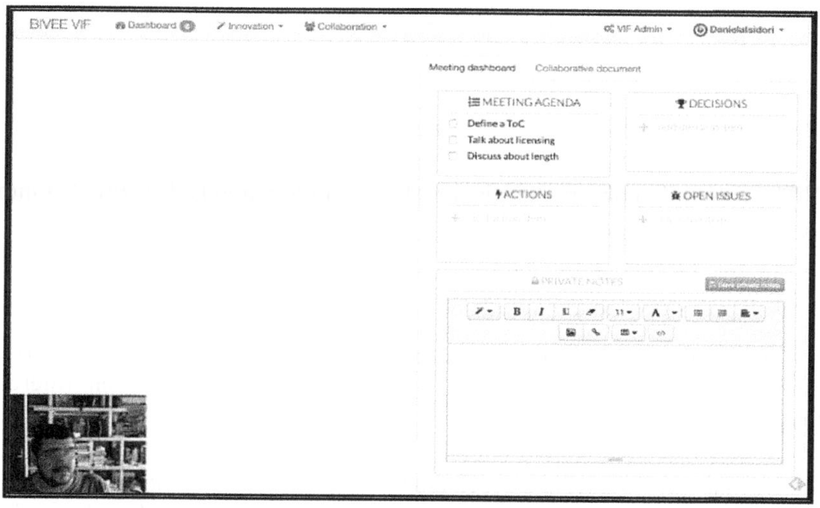

Figure 9.10. *Meeting main screen*

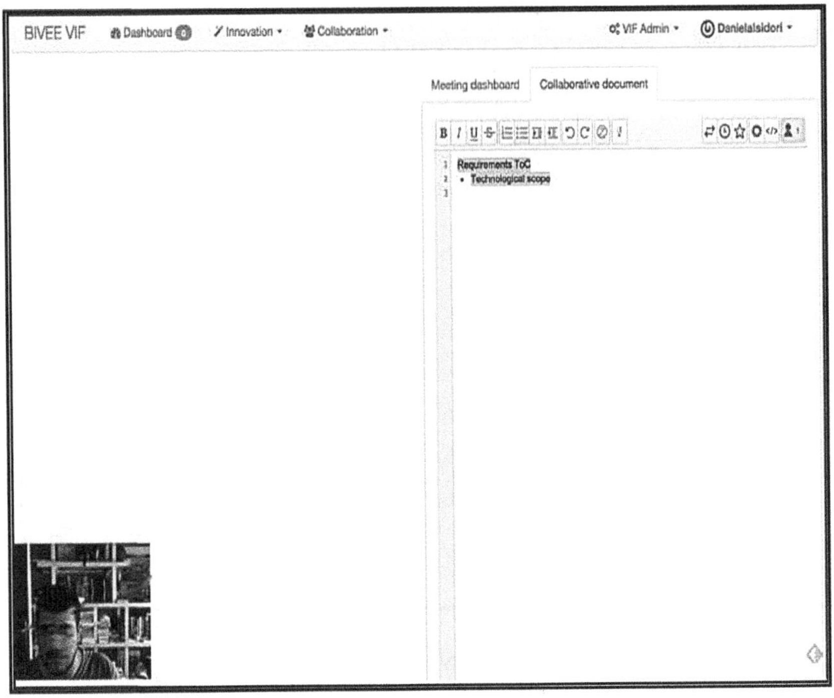

Figure 9.11. *Meeting room, collaborative doc editor*

9.3.9. *The innovation observatory*

The innovation observatory is the place to follow trends. It is divided into two main components:

– the backend process to filter the web based on the provided ontologies. Its results are stored in the VIF database;

– the frontend application to show latest results and to filter those results based on the user queries, and to launch ideas based on filtered elements.

The innovation observatory offers the possibility of following some social networks to create a wall, where all the results are offered and can be analyzed in a more visual way. This component is shill under development, but some results are already shown.

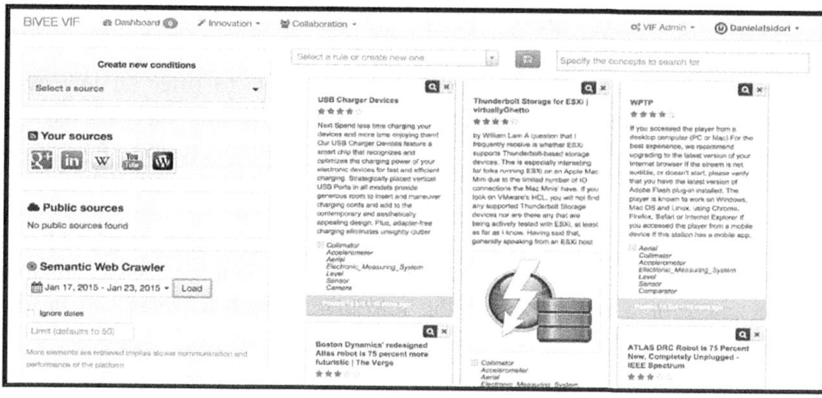

Figure 9.12. *Innovation observatory with results based on ontologies from Twitter*

9.3.10. *The semantic shared whiteboard*

The evolution of this key component has changed from a central tool to a more collaborative-oriented tool.

This semantic shared whiteboard allows the users to create a collaborative and real-time-based whiteboard to add notes, documents, links, etc. Again, this tool is intended to provide visual feedback, which produces better results rather than reading lots of texts.

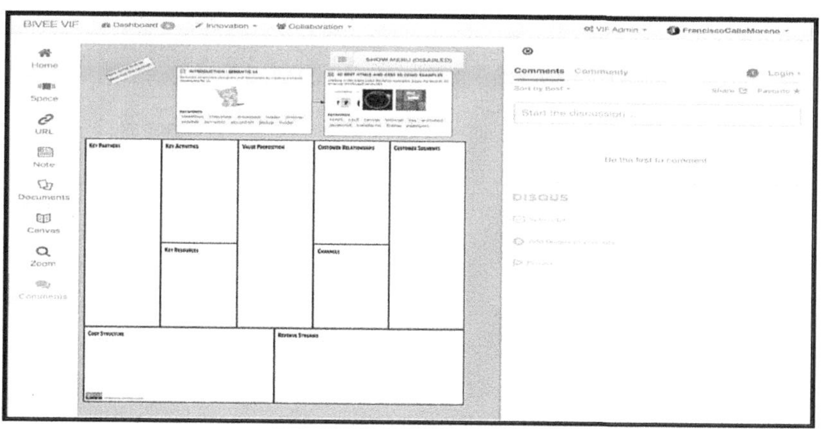

Figure 9.13. *A whiteboard in use*

9.4. Connection with other BIVEE components

The VIF connects with other BIVEE components to offer an improved functionality. The main components in play are the PIKR to store and retrieve knowledge, the RDH to store metrics of the innovation process, the overall platform to provide a single point of access and the mission control room.

The access to the tool is provided by the single sign on process; it allows users from the BIVEE platform to log into the VIF without the need of creating new per-platform users and inheriting information such as roles and profile information.

Interactions with the PIKR provide the VIF with semantic capabilities to annotate, discover and relate documents, ideas, projects and members. Those annotations can be performed fully manually from the UI provided by the VIF or using the automatic document annotation extraction provided by the PIKR.

The RDH is the interface used by the VIF to store metrics about the usage of the platform and to store KPIs measurements (KPMs). In this last case, the information is also kept into the VIF system to be analyzed and accessed in a faster way.

Last but not least, the interaction with the mission control room was established on how to provide improvement to the production space. From a high level point of view, the process can be described as shown in Figure 9.14. This process includes the common understanding between both systems and continuous feedback from one space to the other.

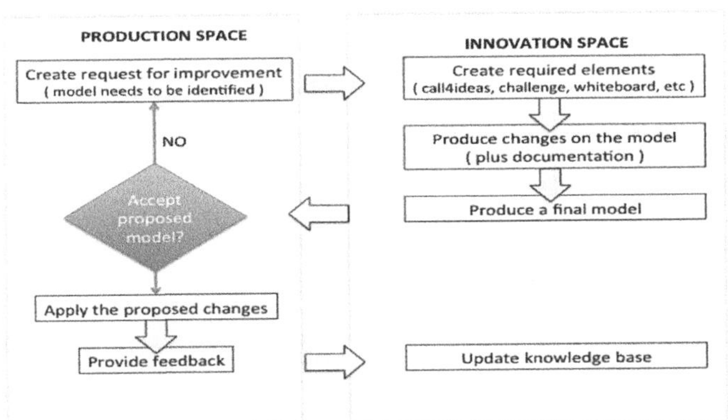

Figure 9.14. *Improvement process*

9.5. Conclusions and future work

After all the work done and the outcomes, the project can be considered as a step forward toward the innovation of the Innovation. The VIF has been focused on the human aspect of the project taking into account the innovative approach provided by the BIRF and mixing it with cutting-edge technologies creating a set of tools that can be the base for future tools and that can help CIOs in establishing a culture of innovation within their enterprises (virtual or not).

The tools created for the innovation space may not be ready for the market at the moment of writing this book, but they are no far away from that. The roadmap to obtain that status includes steps to improve the stability and the UI plus refinement in the integration with other BIVEE components. Due to the modular approach used and, since the technologies used for the current VIF are flexible enough to connect with third-party software – both legacy system and new ones – the VIF can easily be adapted to new requirements in terms of functionality or technology evolution.

The VIF is an open source project and is expected to grow in terms of stability and offered functionality; the current version of the VIF has yet to be considered a prototype. Atos is particularly interested in continuing the development of the VIF for several reasons: to use the VIF as a baseline for possible future European projects and to become an asset to be used internally and externally, when offered to potential clients.

The platform needs to evolve in two ways: first, it needs to focus more on the encouragement of creativity and to foster that creativity toward a better innovation. Second, it has to evolve in a technological dimension, in line with the current trends to make the platform more attractive to the users.

Finally, the outcomes of European Projects, as BIVEE, need to be followed so that the retrieved knowledge can be applied elsewhere. This may lead to a more up-to-date baseline and create the path for future innovations.

9.6. Bibliography

[DAR 12] DARROW B., Meteor rakes in $11.2M to fuel enterprise app development push, available at https://gigaom.com/2012/07/25/meteor-rakes-in-11-2m-to-fuel-enterprise-app-development-push, retrieved 18 January 2015, 2012.

[DBE 13] DB-ENGINES, RDBMS dominate the database market, but NoSQL systems are catching up, available at www.db-engines.com, retrieved 24 November 2013.

[ECM 11] ECMA INTERNATIONAL, ECMAScript® language specification, available at http://www.ecma-international.org/publications/standards/Ecma-262.htm, 2011.

[ECM 12] ECMA INTERNATIONAL, C39 – ECMAScript® (formerly TC39-TG1), available at http://www.ecma-international.org/memento/TC39.htm, 2012.

[ECM 13] ECMA INTERNATIONAL, The JSON data interchange format, available at http://www.ecma-international.org/publications/files/ECMA-ST/ECMA-404.pdf, 2013.

[GIT 14] GITHUB, Inc., DDP specification, available at https://github.com/meteor/meteor/blob/master/packages/ddp/DDP.md, 2014.

[MET 14] METEORHACKS, MongoDBOplog and Meteor, available at https://meteorhacks.com/mongodb-oplog-and-meteor.html, retrieved 22 January 2015, 2014.

[RAM 12] RAMBLINGS B., Fibers and threads in Node.js – what for?, available at https://bjouhier.wordpress.com/2012/03/11/fibers-and-threads-in-node-js-what-for, retrieved 02 February 2015, 2012.

[W3C 14] W3C HTML WORKING GROUP, HTML5 – a vocabulary and associated APIs for HTML and XHTML, available at http://www.w3.org/TR/html5, 2014.

10

The Production and Innovation Knowledge Repository

10.1. Introduction

A *virtual enterprise* is a networked organization where different autonomous entities collaborate toward a common goal. The business innovation in virtual enterprise environment (BIVEE) project introduces the notion of *virtual innovation factory* (VIF) as an innovation-oriented virtual enterprise whose goal is to support the *production* of innovation. Production of innovation takes place in the *business innovation space* (BIS), where active entities, such as the research and development (R&D) innovation teams, final users, partners or even competing companies and other actors, operate. Operational processes pertaining to the BIS define the core innovation value chain: sourcing raw ideas, transforming them into products (goods and services), and marketing and delivering new products. Finally, processes for the planning and monitoring of innovation projects conducted in the BIS characterize the managerial level of a VIF. All these kinds of activities require intensive collaboration, communication and interaction, and ultimately a high level of knowledge sharing among the involved autonomous actors. Indeed, knowledge is the main factor that enables continuous innovation in a world of rapidly changing markets, products, services and technologies (see, for example, [NON 00] for a thorough discussion of the notion of knowledge and its role in enterprise management). Efficient access to knowledge resources is, however, hindered by interoperability issues coming from fragmentation and heterogeneity of the involved players, their data, information and knowledge resources.

Chapter written by Francesco TAGLINO and Fabrizio SMITH.

In order to address interoperability issues and enable sharing of knowledge, the BIVEE project proposes an approach based on *semantic technologies.* In particular, this chapter presents the design principles and the prototype definition of the *Production and Innovation Knowledge Repository* (PIKR), a semantics-based repository for knowledge resources related to the BIS where the VIF operates.

The PIKR is a virtual repository since actual resources (e.g. enterprise documents) physically reside locally, at the premises of the individual virtual enterprise (VE) partners, while the PIKR maintains and manages an *ontology-based* image of such resources, as the result of their *semantic description.* Leveraging on the semantic description of enterprise resources, the PIKR provides a set of services for supporting smart access to stored resources, facilitating the sharing of contents and supporting the information and knowledge interoperability with the ultimate goal of supporting innovation project management.

The design principles of the PIKR descend from the analysis of user requirements and the methodological framework elaborated within the BIVEE project. The analysis led to the recognition of core elements to be semantically described, namely *documents*, *business processes* (BPs) and *key performance indicators* (KPIs), and the relationships among them.

10.1.1. *BIVEE innovation framework*

The BIVEE project elaborated the *business innovation reference framework* (BIRF) [ROS 12] that works as the methodological framework to create and manage knowledge in the BIS. The BIRF is characterized by the notion of innovation waves, in which an innovation venture is articulated, and in which the activities carried out during the innovation lifecycle are organized. Four waves have been identified:

– *Creativity:* all the activities related to the creation and first evaluation of new ideas.

– *Feasibility:* the scope and the intended impact of new ideas are defined, including a first account of their technical and financial feasibility.

– *Prototyping:* the first implementation of the evolving ideas, achieving a first full-scale working model. Such a model is tested and analyzed to verify the actual performance and features, also giving the possibility to rethink some design.

– *Engineering:* activities aimed at producing the specification of the final version of the new product (essentially the bill of materials and manufacturing procedures), ready for the market, and the corresponding production process.

For each wave, a number of activities to be carried out in order for a wave to be completed have been identified. Examples are idea generation and idea analysis for the creativity wave, resource analysis and feasibility study for the feasibility wave, resource allocation for the prototyping wave, build and optimize for the engineering wave.

Each activity is further specified by providing the description of inputs and outputs in terms of knowledge chunks needed and produced by the activity, respectively, that will correspond to a document to be filled, in order for each activity to be completed. Finally, activities and documents are associated with a set of indicators expressing the quality of a document or the performance of an activity. An example of a document related to the (output of) idea generation activity is "Proposed Idea", while the activity can be evaluated by indicators such as the *cycle time* and the *investment in employee development*. The systematic organization of indicators and documents includes the definition of indicators categories and objectives, documents prerequisites and constraints, and so forth.

10.1.2. *Analysis of requirements*

The starting point of the PIKR specification is the analysis of requirements that arise from the BIRF, discussed above, whose basic elements have to be semantically described, and the functional requirement provided by the BIVEE end-users. From the end-users' point of view, the PIKR should enable managers to perform the following activities: (1) to evaluate ideas that are proposed within the VIF, to filter them, and hence to support the selection of promising ideas; (2) to monitor the status of an innovation project and its evolution; (3) to classify ideas; (4) to discover similarities among ideas at any stage (i.e. wave) of their development, for avoiding duplication, and improving the overall efficiency of the innovation process; (5) to reuse experience from past projects, by looking at results of both successful and dead projects and (6) to search for partners with specific competencies.

A selected list of the identified functional requirements follows: (1) to take into account different kinds of enterprise *knowledge resources*, such as documents, processes, products and KPIs; (2) to allow the management of VIF knowledge resources by means of *classification* and *annotation mechanisms*, taking into consideration application domain content of knowledge resources; (3) to support *access* and *retrieval* of resources by means of advanced search functions, based on exact and similarity matching; (4) to keep track of *links* and *dependencies* among knowledge resources for a smart access and manipulation of knowledge resources; (5) to support the *evaluation of indicators* for assessing the status of the activities in a VIF; (6) to provide *reasoning capabilities* over enterprise knowledge resources and (7) to provide *collaborative mechanisms* for facilitating the participation of the largest number of players in the process of continuous optimization and innovation.

The first two requirements are addressed by the internal organization of the PIKR, by means of an ontology-based framework that allows the semantic representation of different kinds of knowledge resources (see section 10.3). The rest of the requirements list is covered by semantics-based search functions and reasoning capabilities.

The remainder of the chapter is organized as follows. Section 10.2 introduces the key enabling technologies adopted by the PIKR. Section 10.3 describes the PIKR ontological framework that enables the semantics-based unified view given by the PIR in a virtual enterprise scenario. Section 10.4 presents the developed methodology for building the application domain ontologies. Sections 10.5 and 10.6 report on the techniques for the semantic annotation of the different kinds of knowledge resources. Section 10.7 outlines the implementation of the PIKR. Finally, section 10.8 concludes the chapter.

10.2. Key enabling semantic technologies

While the BIVEE platform, through its *front-end*, provides several tools (such as whiteboards, serious games and rewarding mechanisms) to foster social interactions and collaborative activities for supporting open innovation tasks, the PIKR offers to such tools services that enable the organization, interoperability, sharing and smart access to knowledge resources. The proposed infrastructure provides the following baseline services:

– *Ontology-based informative resource representation*: the PIKR is organized as a federation of reference ontologies, which are primarily used to semantically enrich any kind of tangible and intangible artifact in the scope of a given VE. In particular, the ontologies are partitioned into knowledge resource ontologies (KROs), which are independent of any application domain and provide the means for the representation of the main knowledge resources (e.g. members' profiles/competencies, BPs, reports and performance indicators), and domain specific ontologies (DSOs) which provide the semantics of a specific business scenario (e.g. automotive and cultural heritage).

– *Collaborative knowledge engineering*: the PIKR supports the building of domain ontologies shared and agreed within the VE organizations. An iterative and incremental process, supported by a software platform built upon a semantic wiki, allows a community of practice, including knowledge engineers, domain experts and ontology stakeholders to cooperate for: (1) producing conceptual models and reaching a consensus on their suitability with respect to the application domain at hand, and (2) guaranteeing at the same time a sound formal encoding into a computational ontology. Depending on the particular requirements at hand, the output can be a full-fledged axiomatic theory, encoded as an OWL (Ontology Web Language) ontology, or a rich semantic network, implemented according to the simple knowledge organization system (SKOS) standard.

– *Semantic access and processing of knowledge*: the PIKR exposes a set of services that exploit the ontology-based knowledge representation, in order to provide advanced semantic facilities for data harmonization, search and retrieval, and verification of business rules. Among them, a similarity engine, implementing the *SemSim* semantic similarity method [FOR 13], enables (1) concept-driven search service, (2) recommendation of contents to users, by matching users' profiles with annotated resources and (3) correlation of knowledge fragments based on their semantic affinity.

10.3. Ontological framework

The objective of the PIKR is to create a semantics-based unified view of the information and knowledge created and transformed within the BIS where the VIF operates, by providing a semantics-based infrastructure for: (1) the representation of *digital documental resources* (DDRs) in terms of shared ontologies and (2) the enactment of reasoning and searching functionalities on the annotated DDRs.

We identified the following core elements to be semantically described. *Documents*, which are concrete footprints of all kinds of activities, both at production and innovation level; *BPs*, which describe all the activities related to the development of innovation projects and their implementation in production processes; *KPIs* for monitoring the progress of innovation projects and the related operational activities. Besides the aforementioned core elements identified in the BIVEE business innovation reference framework, knowledge related to the specific application domain is taken into account in terms of actors and their competencies (which refer to the capabilities of the VIF and its members) and business-related information, such as bill of materials and other production constraints, production techniques, patents and so forth.

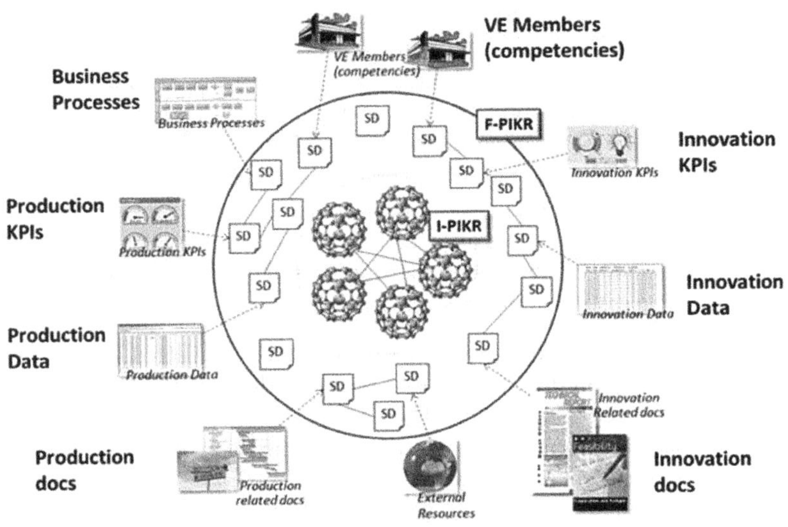

Figure 10.1. *The PIKR structure*

To this end, the structure of the PIKR is organized into two semantic layers (see Figure 10.1): (1) a federation of ontologies to deal with the aforementioned kinds of knowledge, collectively referred to as the *intensional PIKR* (I-PIKR), and (2) the semantic representation of the information relevant to VIF members in terms of the I-PIKR ontologies, through semantic descriptors (SDs), which constitute the *factual PIKR* (F-PIKR).

10.3.1. *Knowledge resource ontologies*

The enterprises that participate in a VE have different methods and tools to represent information and knowledge. The PIKR has the primary objective of acting as a common hub for the management of knowledge resources in a

VE. In this perspective, an important role is played by the KROs, which are briefly introduced here.

10.3.1.1. *DocOnto: document ontology*

In virtual enterprises, document transfer is one of the commonly used ways to exchange business knowledge among parties. The document ontology (*DocOnto*) provides the means for the semantic categorization and annotation of "documents". Running through the four innovation waves, many documents are produced, used, consumed and evaluated. For instance, in the *creativity wave*, given a problem or issue, many ideas can be proposed to address it. Some of them will pass the initial stage and will be further elaborated; some other will be discarded for different reasons. For instance, an idea could be discarded because it appears not to fit the current market conditions, or because the existing technologies are considered not mature or too expensive for gaining a valuable profit. Being able to record such information allows people to keep track of reasons that have guided decisions, and to represent a means for reusing acquired knowledge to save time and money in the future, if similar initiatives are proposed. For instance, in elaborating an idea, to dispose of documents produced in the previous project experiences that share the same or similar topics (e.g. the addressed research line, as well as the used technologies) of the current project results as relevant knowledge resources.

Starting from the analysis requirements from the BIVEE pilot users [SIN 12], and the BIRF, a set of *Infosets* (types of documents) for each innovation wave has been identified as shown below:

– *creativity*: proposed idea, technical solution, preliminary market analysis, innovation report, prospective budget and creativity assessment;

– *feasibility*: project proposal, alliance, market analysis, periodical subproject report and feasibility assessment;

– *prototyping*: requirements, implementation roadmap, prototype technical report and prototype assessment;

– *engineering*: protocols, bill of materials, budget, quality report, test, product data sheet and engineering assessment.

InfoSets are organized into three main sections grouping different kinds of *InfoItems*, which represent smaller pieces of information (i.e. properties):

– *header* groups *InfoItems* such as the title of the document (or part of it), an abstract, the authors and contributors and the URI of the concrete document to be used for retrieving it;

– *content* provides a definition of the content explicated by an informative resource in terms of the DSOs;

– *related knowledge resources* allow *InfoSets* to be related between them (e.g. an *AssessmentReport* can be linked to the *InfoSet* where evaluated contents are described, e.g. a *ProposedIdea*).

10.3.1.2. *ProcOnto: business process ontology*

The business process ontology (*ProcOnto*) provides the means for representing and reasoning with process knowledge, by semantically describing BP models, and makes them available through reasoning facilities. SDs defined according to the *ProcOnto* allow us to achieve a machine-processable representation of a BP model, which can be: (1) enriched in terms of PIKR ontologies, to capture the knowledge regarding the entities involved in such a process, i.e. the business environment in which BPs are carried, and (2) related to other information resources through links among SDs.

For the definition of the ProcOnto, we propose to adopt the business process abstract language (BPAL) [MIS 11, SMI 13], a process ontology which provides constructs for capturing the structure and behavior of a BP represented according to a workflow perspective. BPAL supports the definition of a BP schema, by means of constructs to represent control flow dependencies, data flow, hierarchical decomposition of activities, preconditions and effects of activities. Thus, it is equipped with a formal behavioral semantics, which models the traces that are produced by the execution of BPs. The behavioral semantics of a BP schema captures the notions of state of the world and state updates, caused by the execution of activities, which reflect possible enactments of the BP (i.e. execution traces).

The *ProcOnto* thus encompasses modeling notions common to the most used and widely accepted BP modeling languages and, in particular, its core is based on BPMN 2.0 specification [OMG 11]. Being the constructs provided by the language a sufficient means for capturing structural and behavioral properties of a significant fragment of the BPMN standard, they allow us to deal with a large class of process models.

In the following, we introduce the main aspects of the *ProcOnto* by means of an exemplary *SD* (Table 10.1), resulting from the annotation of a production BP.

bpal: Activity	
Header	
Title	Delivery by external agency
Description	Arranging a goods delivering performed by an external retailer
Identifier	Delivering
Source	http://bivee.eng/ps/aidima/bp_mod_145.uml#Delivery_by_external_agency
Coverage	Production Space
ConformsTo	UML Activity Diagrams
Content	
TermRef	Shipment AND Outsourcing
Condition	ApprovedOrder(po) AND AllocatedInventory(po,in) AND DeliveryDeptUnavailable(po)
Effect	FulfilledOrder(po) AND IssuedInvoice(po,i)
Related Resources	
Participants	controller → SalesDept, performer → Retailer
Control Flow Dependencies	predecessor → analysis_order_size
Data Flow Dependencies	input → DeliveryOrder, output → Invoice, output → PackingSlip
Performance Indicator	reliability →DeliveryPerformance

Table 10.1. *Excerpt of semantic descriptor for the activity "Delivery by external agency"*

The descriptor reported in Table 10.1 is an instantiation of the *ProcOnto* class *activity*, representing a unit of work performed within a BP for the "Delivery by external agency". In the header section, a number of meta-data are collected: Title (i.e. a label to refer to the activity), Description (i.e. a natural language definition of the activity), Identifier (i.e. the system unique reference), Source (i.e. the URL where the resource is published, which in this case is a fragment of a BP model made available by AIDIMA), Coverage (i.e. the scope of the activity, which in this case is the Production Space), Conforms To (i.e. the language the BP model is conformant to), which in this case is unified modeling language (UML).

The content section of a *ProcOnto activity* is specified in terms of: (1) a *terminological annotation*, which defines a correspondence between elements of a BP schema and concepts defined in the DSOs; in the example, the activity is described as a shipment which is given in outsourcing, (2) the *condition* under which it can be correctly executed, which in the example is represented by a situation where items in the inventory have been allocated to fulfill an approved order, if the delivery department is not available, and (3) the *effects* on the state of the world upon its execution, which in the example include the issuing of an invoice.

Other resources that can be related to an *SD* of an activity encompass: (1) the participants to the activity, e.g. the sales department assigned to the supervision or the retailer which actually performs the activity; (2) the control flow dependencies, e.g. the fact that the analysis of order size precedes the delivering; (3) the data flow dependencies, describing inputs (e.g. the delivery order) and outputs (e.g. packing slip) of the activity and (4) the KPIs associated with the activity for monitoring, e.g. the delivery performances to measure the activity reliability.

10.3.1.3. *KPIOnto: key performance indicator ontology*

Several definitions of performance indicators, different for goals, domain of interest, degree of precision and formalism, are provided, e.g. by international and national public bodies, in reference models such as the supply-chain reference model[1] and value reference model (VRM). For the design of the *KPIOnto* and for the development of the BIRF, our approach aims to abstract the general features of such models, with a particular focus on VRM, which explicitly addresses networked enterprises. However, the *KPIOnto* allows us to describe indicators having different scope: "global" indicators, conceived to monitor the whole innovation project, and "local" indicators conceived to measure the performances of individual enterprises. It is worth noting that the latter are needed to derive, namely to compute, the former. For a more comprehensive description of the *KPIOnto*, please refer to Chapter 11.

10.4. Domain ontology building methodology

The conceptualization of domain-specific knowledge, shared and agreed by the VE organizations, and encoded in computational ontologies to be used for capturing in a formal and unambiguous way the knowledge resources of interest, is a crucial aspect. Here, the major challenge is to allow the participation of workers to elicit/codify their tacit knowledge, providing the means to convert it into something more explicit and processable. To this end, in the BIVEE project,

1 http://supply-chain.org/f/SCOR-Overview-Web.pdf.

a collaborative ontology building methodology has been developed. Such a methodology allows a community of practice [IRI 07], including knowledge engineers, domain experts and ontology stakeholders to cooperate for: (1) producing conceptual models and reaching a consensus on their suitability with respect to the application domain at hand, and (2) guaranteeing at the same time a sound formal encoding into a computational ontology.

Figure 10.2 presents an overview of the proposed approach, and its positioning within the socialization, externalization, combination, internalization (SECI) process [NON 00], a reference scheme that addresses knowledge production as a continuous enrichment process based on tacit and explicated knowledge transformations. Socialization is the mechanism of conveying new tacit knowledge in a community of practice through shared experiences (tacit to tacit). Knowledge acquisition and sharing can be facilitated through virtual environments, which are provided in our approach by a (semantic) wiki environment. The externalization that realizes the transformation of knowledge from a tacit to an explicated form is addressed by our approach through iterative phases aiming at an incremental construction of the ontology (i.e. from a lexicon to an axiomatic theory). The organization and integration of knowledge through collaboration and voting facilities ease the combination of different types of explicit knowledge, originating from different individuals. Finally, the reasoning capabilities enabled by a formal representation of knowledge support the ability of people to see connections, recognize patterns and the capacity to correlate fields, ideas and concepts. This contributes to the internalization, i.e. the process of embodying new knowledge in the human mind and social practices (explicit to tacit).

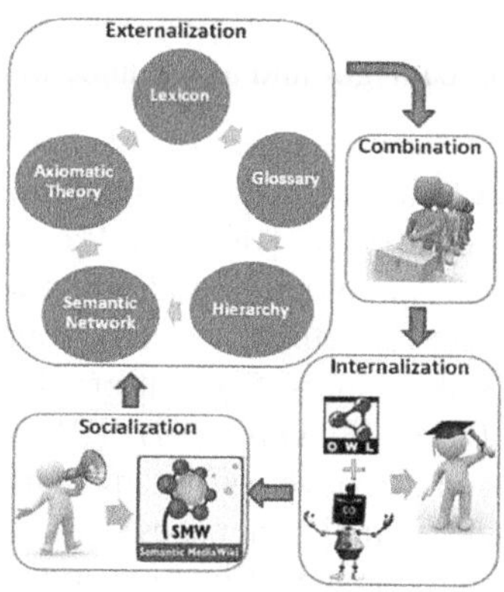

Figure 10.2. *Ontology building methodology overview*

The methodology introduced here provides a collaborative framework for defining, in a progressive way, a business ontology. A number of phases are iteratively conducted, each of which leads to a different outcome, namely lexicon, glossary, hierarchy, semantic network and axiomatic theory. According to this, an initial lexicon is enriched step-by-step with definitions, relations and finally with formal axioms with a level of rigor which depends on the requirements at hand:

– *Targeted knowledge*: the proposed method has been conceived to support the representation of enterprise knowledge in terms of three main axes: (1) the involved players (VE members and potential clients), (2) production and innovation activities in the scope of the VE, associated with, for example, the available capabilities, skills and technologies, and (3) business objects pertinent to the VE reality, including, for example, business objects, documents and products (services and goods). To this end, we introduce some basic conceptual categories (referred to as kinds) following the object process actor language (OPAL) [DAN 07], to support the participants in the conceptualization process, identifying active entities (actors), transformations (processes) and passive entities (objects), with their properties (attributes). Therefore, upper level concepts are: (1) *opal:Actor*, representing active elements of a business domain, able to activate, perform or monitor a process, (2) *opal:Process*, representing any business activity or operation aimed at satisfying a business goal and operating on a set of business objects, (3) *opal:Object*, to represent entities a BP operates on, and (4) *opal:Attribute,* representing a measurable property associated with an individual (e.g. weight, name and color). We adopted OPAL as a "default" component of the proposed framework since it is a light-weight language, which has been validated in several international projects exhibiting a good trade-off between ontological rigor and acceptance in business realities.

– *Involved participants*: communities of practice have been identified as the place where the conversion of tacit and human capital into explicit and structured social capital can occur [IRI 07]. We assume a community of practice involved in an ontology building process composed of a heterogeneous set of people, which we can roughly group as content providers and ontology stakeholders. Content providers can be distinguished as *knowledge engineers* (KEs), *domain experts* (DEs) and *application experts* (AEs). KEs are supposed to have skills in semantic technologies (e.g. conceptual modeling and formal languages), whereas DEs and AEs are experts, respectively, of the domain of interest and the particular application addressed. Among them, specific roles are entitled: the *ontology master*, who is the responsible coordinator of the whole building process; the *mediators*, who lead participants in collaborative activities. Ontology stakeholders can be users, i.e. the group of people who will operationally interact with ontology-based applications or, more in general, people who are interested in the success of the ontology building project.

– *Collaboration means*: each phase of the ontology building process is delimited by deadlines, which are set by the *ontology master*. Within a phase, participants can express their contributions through three collaborative interactions means. Proposing content, which allows content providers to introduce new pieces of knowledge at each phase of the ontology building process (e.g. suggesting a new term, as well as proposing a natural language definition for a given term). Discussing relevant topics, such as: (1) validation issues related to the proposal, acceptance and removal of contents, and (2) strategic aspects related to, for example, the usage and consumption of the ontology, trends of the addressed domain and extensions in the direction of specific/emerging subdomains. Every member of the community can participate to discussions, coordinated by mediators who are in charge of monitoring discussion threads, highlighting the essence of the argumentations, and in case, proposing voting sessions for reaching a final agreement on discussed issues. Voting, which enables all the participants to express their preferences on proposed contents in a poll manner. Votes are expressed according to a graduated rate scale (e.g. 0=discard, 1=neutral and 2=accept) and can be supported with comments. We considered formal techniques [LIN 02] and [DEL 75]. In both, a mediator coordinates the activities without actively taking part to the vote. Furthermore, while in [LIN 02] voting is anonymous, in [DEL 75] it is organized in physical brainstorming sessions where all actors are involved. In our approach, we decided to adopt a non-anonymous voting system. This allows us to identify the most active people in the collaborative process, as well as their effective contribution to the ontology construction. Such an information is exploitable by the ontology master for selecting mediators on the basis of their quantitative, as well as qualitative participation. Voting activities determine an acceptance rate on contents to be validated. Customizable voting policies allow the ontology master to specify: (1) a quorum, which is the minimum number of needed votes to validate a proposed piece of content (e.g. 25% of the participants), as well as (2) a threshold on the sum of the collected votes (e.g. average rate above 60% of the highest allowed rate). Collaborative activities in our approach do not need the organization of physical meetings since they are supported by a web-based infrastructure, which allows remote interactions at any time and from any place.

10.5. Semantic annotation

10.5.1. *Ontology-based lifting of value production space knowledge*

The actors operating in the value production space (VPS) are supported in their work by the BIVEE platform, through the mission control room (MCR) [WOI 12], which provides its facilities through a hybrid meta-modeling platform, whose building blocks include: (1) the *modeling language*, which provides the means for building conceptual models, through a number of constructs predefined according to their semantics, syntax and graphical notation, (2) the *modeling procedures*, which define the stepwise usage of the modeling language, and (3) *mechanisms* and *algorithms* that enable the computer-based processing of models.

Conceptual models, such as BPs, value process chains and e3value models, are the key means for representing enterprise resources to allow stakeholders (e.g. designers, analysts, businessmen and engineers) to discuss requirements, validate design choices and support decision-making. Thus, conceptual models describing the VPS represent the main knowledge asset of the VE, reporting the "as is" situation and facilitating the improvement and the innovation ventures. Nevertheless, the management of such knowledge is limited due to the fact that graphical modeling languages and notations are mainly intended for human consumption, while they fail in making knowledge available in a machine-accessible form. Our objective is to develop a modeling platform that enables stakeholders to manage the knowledge through a human-oriented representation and enrich such representation with machine interpretable semantics for knowledge engineering. The final goal is a hybrid approach aimed at avoiding ambiguities and inconsistencies, providing a unified knowledge space to enable collaboration and knowledge sharing among humans and software applications, and ultimately supporting effective semantics-based production management. In particular, we propose a two-faceted semantics-based intervention. On the one hand, we propose the introduction of a semantically lifted enterprise model, which is represented in a formal, rigorous and machine-processable way, by using ontology-based languages; on the other hand, enterprise models enriched with semantic annotations. This approach enables the implementation of advanced reasoning facilities, for providing smart access and management of stored resources, reasoning over them, facilitating sharing of contents, increasing their quality and supporting knowledge interoperability.

10.5.1.1. *Semantic enrichment of the VPS modeling framework*

For each modeling perspective relevant to the VPS, a modeling language has been defined in the MCR, and ontological reference structures have been formalized in the PIKR, as reported in the following:

– *VE network model*, describing the participating production units, including their organizational structure, capacities and capabilities; to support the annotation of such information, a production unit capability ontology (PUCO) is included in the DSOs, for the representation of actors and related capabilities, as shown in Figure 10.3. In the PUCO, we model *production units* as particular forms of *organizational actors*, which can be aggregated to realize a *VE*. For enriching and standardizing the representation of human actors and organizations, the friend of a friend (FOAF) vocabulary is also included, as shown in Figure 10.3. A production unit offers a number of *capabilities*, described in terms of: (1) a *time interval*, in which the capability can be offered for its implementation, (2) the *KPIs* that the capability is assumed to fulfill, in order to provide a quantitative and qualitative description of the commitment of the production unit (PU) in realizing a given outcome, (3) the *task* related to the capability, e.g. varnishing and delivering, (4) the *business objects* in the scope of the capability, e.g. chair and table, and (5) the *tools/technologies* that a production unit can exploit, e.g. metallic upholstery.

– *Products*: the VPS is centered around the products the VE is able to realize, starting from the bill of materials and decomposing the product in components and finally in raw materials. As discussed in section 10.4, one of the targets of the DSO ontology building process is the object category, including a classification of final products, along with its tangible (*physical*) components (e.g. the monitor of a laptop) and intangible (*non-physical*) components (e.g. the warranty and assistance services associated with a laptop).

– *Business processes*: the representation of BPs in the PIKR is realized through the *ProcOnto*, largely described in section 10.3.1.2.

– *Documents*: the VPS documents in the scope of the BIVEE platform mainly regard documentation at the production unit level, such as guidelines, templates, description of services, service level agreements and the like.

– *KPIs*: the support oriented to monitoring activities given by the PIKR involving the semantic representation and processing of performance indicators is the central subject of Chapter 11.

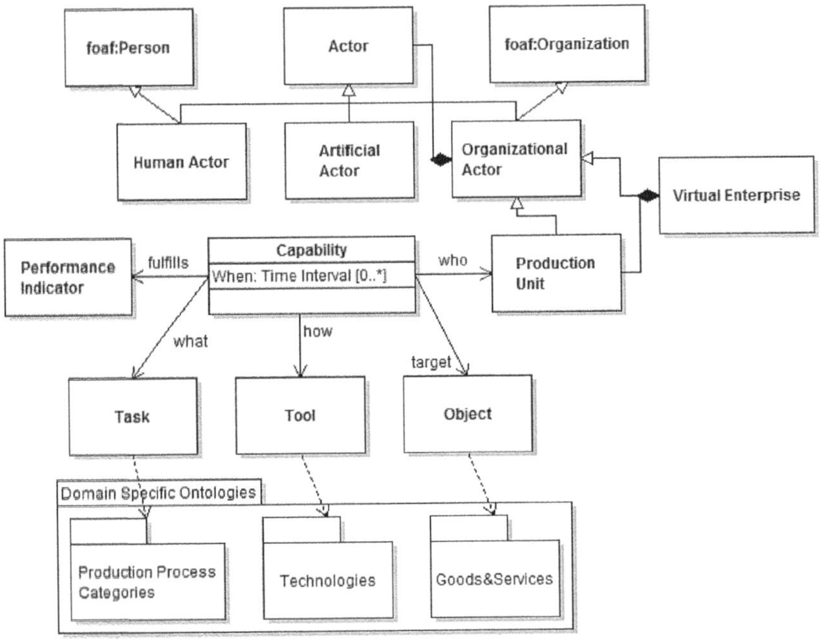

Figure 10.3. *Overview of the production unit capability ontology*

10.5.1.2. *Technical integration of PIKR and MCR*

In order to integrate the semantics-based representation framework realized in the PIKR with the VPS modeling framework realized in the MCR, we follow the semantic transit model proposed in [WOI 13]. This approach requires an extension of MCR meta-model, introducing a semantic model, which enables ontologies to be copied into the MCR using this "transit", in order to be used for the annotation. In order to semantically enrich all the different kinds of modeling artifacts, the integration between PIKR and MCR is based on the following conceptual schema, as shown in Figure 10.4:

– *Meta-model annotation*: the constructs which are defined in the meta-model of a particular modeling language are mapped to concepts defined in a meta-model ontology. Such a mapping can be considered as a predefined built-in of the framework. This kind of annotations relies on the KROs.

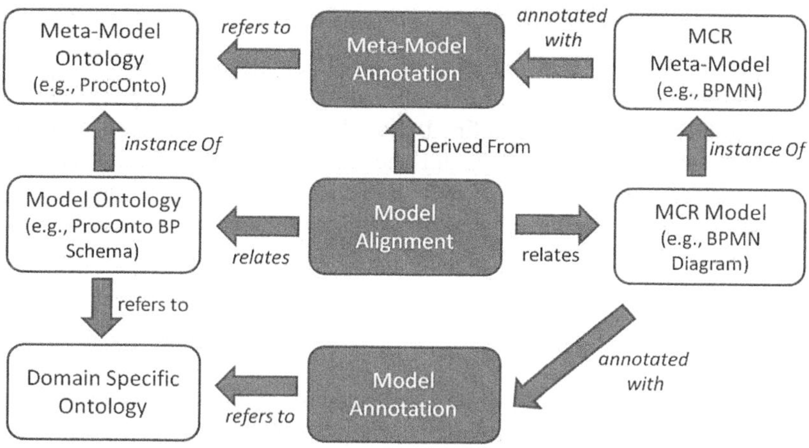

Figure 10.4. *Conceptual integration between PIKR and MCR*

– *Model alignment*: a model alignment includes the correspondences among the entities of a given MCR model, instantiated according to the meta-model *MM,* and a model ontology *MO* instantiated according to a meta-model ontology *MMO*, provided that a meta-model annotation is available between *MM* and *MMO*. Basically, given an MCR model and its annotated meta-model, a model alignment can be automatically derived to generate a corresponding model ontology.

– *Model annotation*: the elements occurring in a model are put in relation to concepts defined in some ontologies; in our framework, model-level annotations rely on DSOs.

The aforementioned integration impacts on the MCR *modeling capabilities* in two ways: (1) semantic models (ontologies) extend the modeling framework and enable the description of modeled entities, by referencing ontology fragments, and (2) automatic transformation of MCR models into interpretable ontologies.

As far as the *mechanisms and algorithms* implemented by the MCR are concerned, we have two main extensions: (1) ontology-based processing of models discussed in [TAG 12] that extends the SQL-based search and retrieval support given by the MCR to search and retrieve facilities discussed in [TAG 12], and (2) knowledge-based evaluation on models, providing semantic compliance checks on a model based on declarative constraints (both syntactically and semantically).

10.5.1.3. *Application scenarios and main functionalities for the VPS*

As far as the VPS is concerned, the BIVEE platform through the MCR provides two main categories of functionalities, related to (1) VPS design, where the business goals are defined, the VE is built, the business activities are structured into BP and KPIs for the VPS are identified, and (2) VPS monitoring and improvement, where KPIs are collected and measured to identify underperforming areas and drive the continuous improvement. Since point (2) is largely discussed in [TAG 13], in the following we focus on applicative facilities regarding VPS setup:

– *VE setup*

Once the goals of the VE to be constituted are defined, together with the requested competencies and production capabilities, the semantic profiling of VE members supports: (1) semantic search of partners for identifying who can take part to the VE, (2) allocation of production units to production activities according to their capabilities, and (3) definition of the bill of material and the identification of the associated production process, by navigating the product decomposition hierarchy.

– *Reuse in modeling*

During the setup of the VPS, the PIKR offers reuse facilities, based on the searching and querying baseline reasoning services. Especially in the VPS modeling phase, a pool of related resources can be suggested by the PIKR. Modeling does not start with a blank sheet, but by formulating a request and selecting the proposed results.

– *Knowledge interoperability*

The adoption of reference structures in the form of KROs and DSOs provides the means for the annotation of heterogeneous knowledge resources in terms of the same language via shared terminological definitions. This implies:

– *Compliance of the designed VPS*

The compliance of the designed VPS with respect to domain-specific constraints can be checked according to the consistency baseline PIKR service. Use cases include: (1) exploiting KPI reasoning facilities for verifying if the measurement data made available by VE members are sufficient or which KPI can be monitored with available data, or guaranteeing that no contradictory indicator will be computed at runtime, (2) verifying VE-specific business rules and regulation against the designed BPs, (3) checking semantic and syntactic correctness criteria formulated for a given modeling language supported by the MCR, and (4) assuring that the documented VE guidelines, description of services and service level agreements are properly reflected in the designed VPS.

10.5.2. *Ontology-based lifting of business innovation space knowledge*

10.5.2.1. *Semantic enrichment of the BIS modeling framework*

Each wave, namely creativity, feasibility, prototyping and engineering, is organized into a certain number of activities to be carried out in order for a wave to be completed. Each activity is characterized by a *set of documents*, which are the artifacts where innovation-related initiatives are accounted. As such, innovation-related documents represent valuable resources for sharing information about ongoing and past innovation initiatives. The main objective of the semantic enrichment of the BIRF is the semantic description of innovation-related documents by semantically abstracting and lifting their structure, and enriching their content in order to enable interoperability and openness, as well as reasoning services such as querying and retrieval of documents, reasoning over documents description, understanding similarities among documents, assessing status and quality of documents, and monitoring innovation activities. The semantic abstraction and lifting is performed in terms of the DocOnto, while the domain knowledge enrichment is performed in terms of the DSOs (see Figure 10.5).

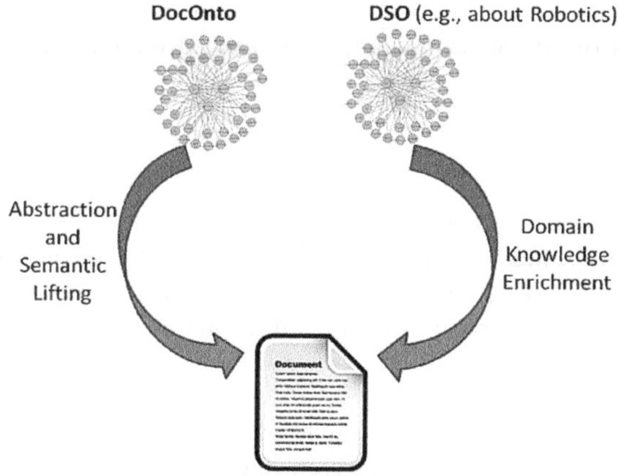

Figure 10.5. *Ontology-based support to document annotation*

In section 10.3.1.1 we reported on the organization of the document ontology (*DocOnto*), in this section we will focus on the technical aspects concerning the interactions between the PIKR and VIF, as well as on the main application scenarios where the semantic enrichment of BIS resources can be exploited.

10.5.2.2. *Technical integration of PIKR and VIF*

The integration between the PIKR and VIF happens by means of a set of adaptors, which enable the VIF to interact with the PIKR for supporting the semantic lifting and the domain knowledge enrichment of documents in the BIS. The adaptors will allow the VIF to: (1) retrieve the structure of innovation-related documents as they are modeled in the DocOnto, (2) retrieve and query the DSOs, and (3) feed the F-PIKR with SDs of documents.

10.5.3. *Application scenarios and main functionalities for the BIS*

Once resources in the BIS (mainly, innovation-related documents) have been semantically described in terms of the DocOnto and DSOs, a number of functionalities can be enacted by the VIF through the interaction with the PIKR services.

10.5.3.1. *Information flow management*

Navigating through the document repository

Following the links defined in the DocOnto and connecting the different types of *InfoSets*, it will be possible to navigate the repository of documents.

Searching for documents

The most straightforward usage of ontology-based description of BIS resources is represented by the possibility of searching for innovation-related documents. As described in [TAG 12], the PIKR provides services for exact and similarity-based retrieval of annotated resources.

Suggesting related documents

While people are accessing a given document through the VIF, related documents could be proposed as additional resources to be inspected.

Proposals can be identified on the basis of: (1) similar contents, for instance given a document annotated as a *ProposedIdea*, other idea documents referring to the same (or similar) *objectives* could be proposed, and (2) dependencies/associations between documents. For instance, given a document annotated as a *ProposedIdea*, the assessment report of the idea itself could be proposed.

Checking consistency and status of the documents in the BIS

The DocOnto defines dependencies between *InfoSets*. For instance, a *ValidatedIdea* report needs to refer to an existing *ProposedIdea*. If such information is not indicated, the system has to identify such an inconsistency and provide feedback to the VIF.

Proposing extensions to DSOs

Since innovation activities can go beyond existing frontiers or at least consider new application domains, it is possible that at the annotation time the DSOs lack the required knowledge. For this reason, the PIKR has to support the VIF in extending the DSOs by proposing a communication channel for proposing new concepts.

Supporting innovation observatory

A semantics-based crawler will support the monitoring of relevant initiatives happening outside the borders of a given virtual enterprise with respect to new innovative initiatives, innovation opportunities and existing initiatives that can be related to the VE's activities.

Semantics-based social media for collaborative open innovation

Posts, comments and discussions of VIF users through the collaborative whiteboard will be semantically annotated/tagged as the other documental resources in the BIS, so that they can be positioned, grouped, localized one next to another and connected to one another forming knowledge clusters, with the intent to create an articulated picture of the hot issues at hand.

10.6. Semantic enrichment of semantic media contents

The analysis of semantic media contents (SMCs) aims at building a semantic representation of all the collective knowledge maintained in the overall VE. To this end, through the BIVEE front-end, SMCs are submitted to the PIKR to be semantically analyzed and annotated.

The annotation is represented by an *ontology feature vector* (*OFV*), i.e. an ordered set of references (URIs) to concepts defined in the reference ontology. The automatically suggested annotation can be validated and manually improved by users.

UC SE1: open collaboration – leveraging on crowd intelligence: when a new issue arises, an instance of the shared whiteboard (SW) is created, and the information related to the case is posted on it. The open sections of the SW are then accessed by *contributors*, which can provide comments and suggestions to find a solution. Each piece of content posted on the SW is automatically annotated by means of the semantic analysis provided by the PIKR. The annotation, expressed as an OFV, can be visualized and refined by each member invited to the SW.

10.6.1. *Semantic search to foster idea creation*

At each stage of the development of an idea, the BIVEE platform offers the possibility of performing a semantic search to automatically retrieve related resources (e.g. past innovation projects, ideas or solutions to problems). As the annotations, search criteria are expressed in the form of a feature vector as well. An autocompletion facility is provided by the front-end for easing the query formulation. As an alternative, focused search requests can be automatically formulated by selecting a given SMC (e.g. a post) and using the associated feature vector as the query to find related knowledge resources.

UC SS1: semantic search on past experience – once a user proposes a new idea through the BIVEE platform, he/she is interested in assessing if similar issues have been addressed in the past and collecting knowledge resources related to these previous initiatives as useful material for a further development of the proposed idea. To this end, the user formulates a search request (through an interactive search widget) by using the annotation associated with the proposed idea as the query, or by manually prompting a set of concept identifiers. The user further refines the target of the semantic search by selecting: (1) the target repository (i.e. the corporate memory of the enterprise, the business ecosystem repository or the outside world), and (2) the kind of resources (e.g. proposed ideas and open discussions) that should be retrieved.

10.6.2. *Semantic correlation of SMCs*

The semantic analysis performed by the PIKR on SMCs includes their indexing, aimed at computing and storing in advance, for each SMC, the clusters of semantically related items it belongs to. This functionality is then exploited to semantically correlate pieces of collective knowledge, as exemplified in the following use cases:

– *UC SC1: knowledge combination in the whiteboard* – a group of users is working on an SW instance. One user triggers the PIKR to suggest automatically related knowledge items. New contents, e.g. open discussions and other SW instances, are added to the SW;

– *UC SC2: duplicated warning – better content advice*: while a user (say U_1) annotates an article (say a_1), posted on an SW instance, another user (say U_2) adds another article (say a_2), on another SW instance, with the same subject of a_1, but referring to a different source. U_2 triggers a request to the PIKR for searching knowledge items semantically related to a_2. During the process of suggesting similar contents, the PIKR *understands* that a_1 and a_2 contain very similar content. This is induced by the equivalence of the two annotations. The notification is raised to U_2, who can: (1) avoid duplication and discard a_2, if a_1 and a_2 represent the same article, and (2) maintain a_2, since it has a high degree of similarity, but a better/richer informative content, with respect to a_1.

10.6.3. *User-driven content browsing*

The following use cases focus on the exploitation of the semantic profile of a user (again given in terms of ontology concepts), and the semantic annotation of SMCs to enhance the navigation experience and the information consumption from a user perspective:

– *UC CB1: semantic routing* – on a regular basis, the platform proactively suggests contents to users on the basis of their profile, through a notification on their personal page or email. Suggestions are computed by the PIKR by retrieving, for each topic in the user profile, the more relevant/recent/rated SMCs. Then, SMCs can be filtered/ranked on the basis of properties such as creation date, documental category, author, etc.;

– *UC CB2: profile-dependent content browsing* – the user browses the contents maintained in the BIVEE platform through the BIVEE front-end, which invokes the PIKR to retrieve resources that are intended to be relevant to the user. To this end, a search request is automatically formulated on the basis of the user profile, and evaluated by the PIKR. Results are returned to the front-end for their visualization;

– *UC CB3: semantic storifying of posts* – the user wants to understand the overall activities/trends regarding a particular issue at a VE level. A search request is formulated given a (set of) topic(s) and a time interval. All the related SMCs posted in the platform regardless authors and sources are arranged along a time line. The user can, for example, visualize such a time line, navigate the links to the original posts, identify groups of people dealing, at a given moment, with similar issues, analyze trends in the conversations and analyze the temporal distribution of participation.

10.7. Implementation

In this section, an updated and revised technical description of the PIKR software prototype is given. The architecture, as well as the technical implementation, has been conceived by following two main concepts: (1) *service-oriented architecture*, in order to have flexible and platform independent technological communication channels among the PIKR and the other software operating in the VE, and (2) *linked data principles*, for exposing, sharing and connecting pieces of data, information and knowledge by using semantic web technologies and standards. In particular, following the linked data approach, the PIKR provides, on the one hand a set of reference structures (i.e. ontologies) for the semantic description of enterprise knowledge resources, and on the other hand semantics-based services for accessing and reasoning about such descriptions. To enforce the openness of the platform from a technical perspective, every knowledge fragment is identified by a URI, accessible via HTTP, described by a resource description framework (RDF)/OWL and processable by reasoning services exposed as Web services.

Figure 10.6 shows the overall PIKR architecture, and its positioning within the BIVEE platform. The PIKR is intended to provide semantic facilities to the BIVEE front-end, which implements the functionalities to assist the innovation team in the management of an innovation project. Such functionalities, implemented by the components shown in the upper part of Figure 10.6, encompass: (1) the *setup of the virtual enterprise*, with the definition of its business goals, capabilities, production plans and so on, (2) definition of the *business models* needed for the enactment of the innovation and production projects (e.g. BPs, objects, goals, requirements and KPIs), (3) the support for the *enactment of an innovation project* along the four waves, for the coordination and planning of the activities conducted by the participant actors, and (4) the *monitoring* of both the VPS and BIS, in order to foster new improvements and further innovations. In doing this, the *front-end* interacts with the *semantic layer*, which offers a semantically enriched and machine-processable representation of the informative resources maintained by the *back-end*. The latter is in charge of (1) acquiring several kinds of resources from the information systems of the VE members through *data adaptors* and make them accessible by the platform through the raw data handler (see [TAG 13]), and (2) storing the different kinds of resources handled by the BIVEE platform (storage system), in particular the documental resources managed by the VIF, the models designed in the MCR and the activity logs relate to both VPS and BIS.

The *PIKR connector* is then responsible for handling the mapping between the data in the storage system and the corresponding semantic representation available in the semantic layer, while the *semantic data handler* (largely discussed in [TAG 13]) enriches the raw data handler to realize the reconciliation and manipulation of the local enterprise data.

10.7.1. *PIKR architecture overview*

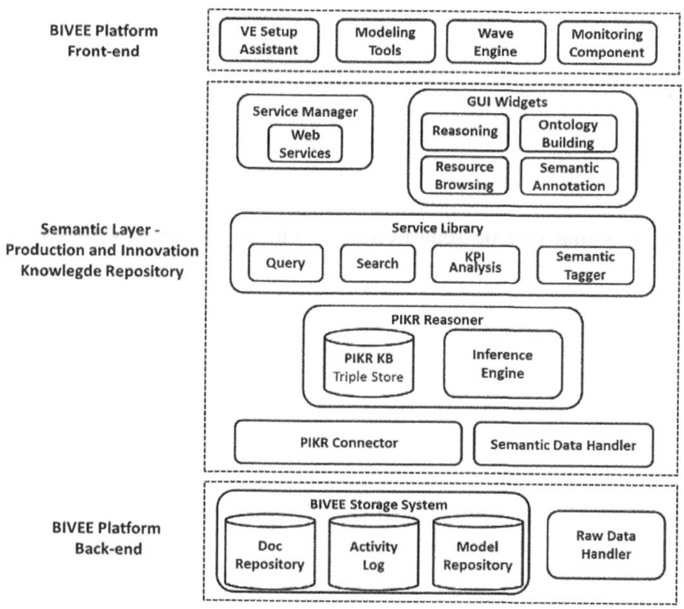

Figure 10.6. *PIKR architecture and positioning within the BIVEE platform*

The entry point for the PIKR reasoning functionalities is the *service library*. Several semantic services are implemented on top of the PIKR reasoner and the semantic resources (SDs and ontologies) constituting the PIKR knowledge base (KB). The latter is stored in the *triple store*, providing basic storage and retrieval facilities for OWL/RDF data. The various reasoning methods that operate on the PIKR KB are implemented by the (rule-based) *inference engine*, and are made available through a Web service interface, exposed by the *service manager*, and through the *PIKR GUI,* which enables user interaction through a wiki-like environment.

10.8. Conclusions

In this chapter, we have presented a semantics-based infrastructure, called PIKR, aimed at supporting innovation-related activities in a virtual enterprise context, by enabling knowledge classification and sharing, as well as interoperability among participating enterprises.

The PIKR has been designed by following the user requirements and the reference framework produced in the context of the BIVEE European project activities. From a technical point of view, the infrastructure is designed according to the linked data principles, describing knowledge resources and their semantic relations in terms of a federation of reference ontologies, and providing entry-points to process the maintained knowledge according to well-known standards (e.g. RDF/OWL and SPARQL) and technologies (e.g. Web services). While the actual knowledge resources (e.g. processes, documents and performance indicators) are stored at the premises of the respective owner companies in the virtual enterprise, the PIKR maintains resource images in the form of SDs that can be regarded as instances of the concepts defined in the ontologies. On top of these descriptions, a set of core semantic services is offered to ease the navigation and retrieval of resources, along with a set of facilities for reasoning about them, intended as a baseline for the development of additional services for populating and exploiting the PIKR.

10.9. Bibliography

[DAN 07] D'ANTONIO F., MISSIKOFF M., TAGLINO F., "Formalizing the OPAL eBusiness ontology design patterns with OWL", *Proceedings of the 3rd International Conference on Interoperability for Enterprise Software and Applications*, Enterprise Interoperability II, Web Services, pp. 345–356, 2007.

[DEL 75] DELBECQ A.L., VAN DE VEN A.H., GUSTAFSON D.H., *Group Techniques for Program Planners*, Scott Foresman and Company, Glenview, IL, 1975.

[FOR 13] FORMICA F., MISSIKOFF M., POURABBAS E. *et al.*, "Semantic search for matching user requests with profiled enterprises", *Computers in Industry*, vol. 64, no. 3, pp. 191–202, April 2013.

[IRI 07] IRICK M.L., "Managing tacit knowledge in organizations", *Journal of Knowledge Management Practice*, vol. 8, no. 3, 2007.

[LIN 02] LINSTONE H.A., TUROFF M. (eds), *The Delphi Method: Techniques and Applications*, Addison–Wesley Publications, Reading, MA, 2002.

[MIS 11] MISSIKOFF M., PROIETTI M., SMITH F., "Querying semantically enriched business processes", in HAMEURLAIN A., LIDDLE S.W., SCHEWE K.D. *et al.* (eds), *22nd International Conference on Database and Expert Systems Applications*, Toulouse, France, 29 August–2 September, LNCS, Springer, vol. 6861, pp. 294–302, 2011.

[NON 00] NONAKA I., TOYAMA R., KONNO N., "SECI, Ba and leadership: a unified model of dynamic knowledge creation", *Long Range Planning*, vol. 33, no. 1, pp. 5–34, 2000.

[OMG 11] OMG, Business process model and notation, available at http://www.omg.org/spec/BPMN/2.0, 2011.

[ROS 12] ROSSI A., KNOKE B., EFENDIOGLU N. *et al.*, Specification of business innovation reference framework (in the context of the VEMF), Deliverable D2.2, BIVEE European Project, 2012.

[SIN 12] SINACI A., PIERSANTELLI M., CRISTALLI C. *et al.*, "A document centric approach for user requirements in BIVEE", *CEUR Workshop Proceedings*, vol. 864, Article 5, 2012.

[SMI 13] SMITH F., PROIETTI M., "Rule-based behavioural reasoning on semantic business processes", *Proceedings of the 5th International Conference on Agents and Artificial Intelligence (ICAART)*, 2013.

[TAG 12] TAGLINO F., SMITH F., PROIETTI M. *et al.*, Production and innovation knowledge repository, Deliverable D5.1, BIVEE European project, 2012.

[TAG 13] TAGLINO F., SMITH F., DIAMANTINI C. *et al.*, Advanced interoperability services, Deliverable D5.3, BIVEE European project, 2013.

[WOI 12] WOITSCH R., EFENDIOGLU N., KNOKE B. *et al.*, State of the art and mission control room specification, Deliverable D3.1, BIVEE European project, 2012.

[WOI 13] WOITSCH R., *Hybrid Modeling with ADOxx: Virtual Enterprise Interoperability using Meta Models, NGEBIS13*, Springer, 2013.

11

Monitoring Innovation and Production Improvement

11.1. Introduction

Today's dynamic businesses recognize the need to monitor their business performance and to proactively make strategic and tactical decisions rather than becoming reactive to situations. Monitoring provides a valid support to detect problems, prevent undesired situations, avoid repeating mistakes, as well as identifying virtuous behaviors both in daily production activities and innovation-oriented initiatives. Key performance indicators (KPIs) are metrics that provide quantifiable data to assess how organizations, business units or individuals are performing against predefined goals and target. KPIs are critical components, which help to bridge the gap between strategy and execution, and are basically classified into two categories:

– *lead* indicators, which measure the activities that have significant impact on future performance (e.g. customer satisfaction scores);

– *lag* indicators, which measure the output of past activities (e.g. most of the financial metrics).

The predictive capability of KPIs depends on the history data that have been collected, and one of the biggest challenges is the consolidation of performance data from disparate sources into a *coherent* system [MAS 07]. Due to the lack of enforced integrity and relationship, disparate data are usually of low quality, create confusion and inefficiencies, and blur accountability. Data integration suffers from heterogeneity issues, which are

---

Chapter written by Claudia DIAMANTINI, Domenico POTENA and Emanuele STORTI.

often related to different terminology, structure, granularity and typing of data.

Such a problem is increasingly critical in a virtual enterprise (VE) environment, where the data providers can be very different and heterogeneous organizations characterized by high levels of autonomy. This problem needs solutions for:

– collecting data from different sources without constraining VE members to change their information schema;

– enabling a uniform view on such data so that data coming from different and heterogeneous sources can be integrated;

– providing smart mechanisms to access and reason over data in order to get data in a selective and intelligent manner.

To address such issues, we propose an integrated suite of semantics-based services to support acquisition, integration of and access to data. At the basis of the proposed solution, there is the need to allow organizations involved in a VE environment to continue to use their information systems, without imposing them any change in the organization and structuring of their data. The proposed approach to data interoperability described hereby is in line with solutions for data integration of autonomous sources in a federated scenario [LEN 02, GOL 12], and is based on the mediation of a KPI reference ontology (KPIOnto) to overcome heterogeneities among data to be used for evaluating performance indicators. By using the KPIOnto, the information elements that a system shares and makes available to the VE environment are semantically annotated in order to represent their intended meaning in terms of the common reference. The query to access the actual data can be expressed in terms of the KPIOnto, and then can be rewritten in terms of the original sources.

The proposed solution is articulated into two steps:

– the first step encompasses *normalization* and semantic annotation activities on the acquired data, in terms of the KPIOnto, in order to resolve naming and typing heterogeneities. This activity is performed once, at the time of the data acquisition;

– the second step is represented by the rewriting of queries, which assumes that data are provided with respect to a common vocabulary. Queries are written in terms of the KPIOnto, and reasoning over the KPI formulas and dimensions allows us to overcome the further data heterogeneities (e.g. structuring and granularities).

The chapter is structured as follows: in the next section, some relevant works in data integration and performance measurement are described. In section 11.3, the architecture of the solution is introduced, and then is detailed in sections 11.4–11.6 as a suite of semantic-based tools and services. The graphical user interfaces of the platform are presented in section 11.7. Finally, in section 11.8, some conclusive remarks are given and we discuss possible future work.

11.2. Related work

Only a few works in the literature have focused on models and methodologies for monitoring efficiency and effectiveness of innovation activities and for innovation project planning, especially in cooperative environments. Many studies propose innovation indicators at macro-level, suitable for analyzing innovation performances of countries or market segments (see, e.g., OECD[1]) but are largely inapt to monitor and improve internal processes and organizational structures. In [ADA 06], a review of the literature pertaining to the measurement of innovation management at the firm's level is presented, reporting also "an absence of measures well aligned to the activities of the innovation process". To the best of our knowledge, there is a lack of work about *reconciliation of heterogeneous KPIs* used by different enterprises in a collaborative setting, which requires a peculiar care in the integration.

Recently, due to the explosive growth of distributed applications, the need of semantic integration has become increasingly critical. In fact, semantic heterogeneity may persist even if both syntactic and schemas heterogeneities do not occur (e.g. naming the same concept differently). *Semantic data integration* relies on conceptual and machine-understandable representation of data and their relationships to avoid heterogeneities and allow interoperability. The main methodologies used in the literature refer to a single ontology approach in which all source schemas are mapped to a central ontology, or a multiple-ontology approach, with mapping among the local schemas and no shared global view, closely related to Global-As-View and Local-As-View approaches [CRU 05]. Ontologies are useful not only as a global conceptualization, but also to support the definition of semantic mappings between the local sources and the global ontology, and to support

1 OECD, *Glossary for OECD composite leading indicators*, see http://www.oecd.org/document/1/0,3746,en 2649 34349 32694145 1 1 1 1,00.html.

the user in the formulation of high-level queries over the global schema and its rewriting into local queries.

Integration is far more complex when the data models of the sources are multidimensional, as in the case of the present work and in the data warehouse field. *Multidimensional models* are specifically suitable to support data analysis rather than perform online transactions, and categorize data as facts with associated quantitative measures, or as a hierarchy of dimensions that describe the facts. The multidimensional model takes into account the aggregative aspect, defining a data cube as a multi-level, multidimensional database with aggregate data at multiple granularities [LAK 03]. The definition of powerful online analytical processing (OLAP) operators such as drill-down directly comes from this model. Semantic representations of the multidimensional model have been recently proposed [HUA 07, LAK 03] mainly with the aim to reduce the gap between the high-level business view of indicators and the technical view of data cubes, to simplify and automatize the main steps in design and analysis [XIE 07] and to improve OLAP functionalities [PRI 03].

Although the complex nature of indicators is well known, the *compound nature of indicators* is far less explored. Proposals in [PED 09, XIE 07, KEH 09, GOL 12, HOR 14, POP 10, DEL 10] include in the representations of indicator properties some notion of formula in order to support the automatic generation of customized data marts, the calculation of indicators [XIE 07, PED 09, HOR 14], or the interoperability of heterogeneous and autonomous data warehouses [GOL 12]. In the above proposals, formula representation does not rely on logic-based languages; hence, reasoning is limited to formula evaluation by *ad hoc* modules. No inference mechanism and formula manipulation is enabled. Formal, logic-based representations of dependencies among indicators are proposed in [POP 10] and [DEL 10]. These properties are represented by logical predicates and reasoning allows us to infer implicit dependencies among indicators useful for organization modeling, design, as well as reuse, exchange and alignment of business knowledge.

An *ontological representation of indicator formulas* is proposed in [KEH 09] in order to exchange business calculation definitions and infer their availability on a given data mart through semantic reasoning. These approaches are mostly targeted at single enterprises. Despite the increasing interest in collaboration and networking, there is still a lack of work in addressing data cubes in distributed and collaborative contexts. A proposal in this respect is [GOL 12], where interoperability of heterogeneous and autonomous data warehouses is taken into account. However, although a preliminary attempt to include formulas in the framework, the compositional semantics of KPIs is not represented.

11.3. Architecture of the performance monitoring

The BIVEE project relies on a knowledge-centric approach in which the semantic layer, namely the production and innovation knowledge repository (PIKR) [DIA 13], represents the foundation supporting advanced functions of the whole environment. It maintains semantic meta-data for all the entities of the VE (from actors to products, and from innovative ideas to production plans) and offers advanced services. In this section, we briefly discuss the back-end modules of BIVEE infrastructures that are more relevant in the context of this chapter, as shown also in Figure 11.1:

1) KPIOnto (described in section 11.4) is an ontology devoted to model KPIs and their formulas, in a global and shared formal model, that serves as a reference for the whole VE. Such a model is useful to define a common vocabulary that is used to semantically enrich enterprise data. By exploiting the ontology, a set of semantic services has been developed with the aim to reason and manipulate on KPI's formulas, as described in section 11.5.

2) The raw data handler (RDH) is the component responsible for gathering enterprise data, for their semantic enrichment and storage as historical KPI data in the data storage of the BIVEE platform, as explained in Chapter 7 and in Chapter 12 in details. Each indicator is represented as a unique sparse and non-uniform cube, i.e. containing many empty cells at different granularity levels. Cubes representing the same KPI from different enterprises are merged together, reconciling dimensional heterogeneities by referring to the ontological representation of dimension of the KPIOnto.

3) The semantic data handler (SDH) (discussed in section 11.6) is the module that allows the other BIVEE components (mainly the mission control room) to access KPI data by providing a uniform view over them, independently of their original data sources. The SDH accepts requests in terms of the VE common reference, i.e. the KPIOnto, rewrites and resolves such requests as queries over the data storage.

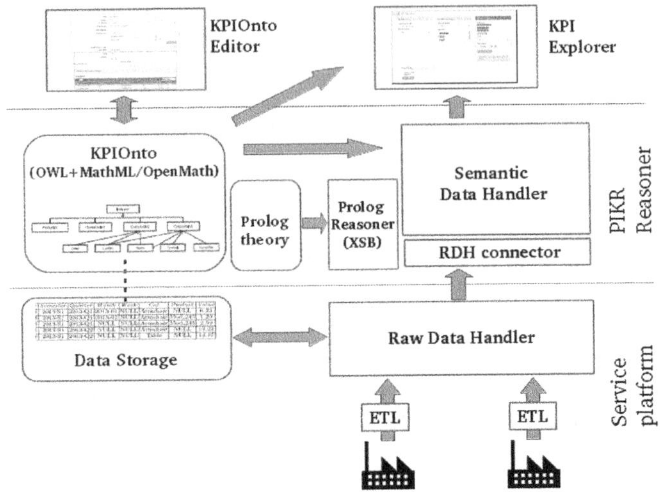

Figure 11.1. *Architecture of the performance monitoring*

11.4. KPIOnto

11.4.1. *Analysis of requirements and KPI characteristics*

There is a plethora of performance indicators definitions and/or glossaries provided by researchers and research groups, international and national public bodies (e.g. OECD) and in reference models like the supply-chain reference model (SCOR)[2], the value chain reference model (VRM)[3] and Six-Sigma [TEN 01]. They all provide quite diverse indicators, depending on the goal and the domain of interest. Almost all proposals are homogeneous in the kinds of information provided, differing on the degree of precision and formality of the description. For the characteristics of accuracy of indicators description in the VRM model, the fact that it explicitly addresses networked enterprises, and since the Business Innovation Reference Framework of the BIVEE project is focused on it, in the following we will mainly focus on the description of indicators in VRM.

2 Supply-Chain Council, *Supply chain operation reference (scor) model*, http://supply-chain.org/f/scor-overview-web.pdf.

3 Value-Chain Group, *Value reference model (vrm)*, http://www.valuechain.org/en/cms/1960.

11.4.2. *Ontology schema*

For such reasons, the main classes are Indicator, BusinessObject, Dimension and Formula. Figure 11.2 shows these categories and the relationships among them.

Indicator: it is the pivotal class of the KPIOnto, and its instances (i.e. indicators) describe the metrics enabling performance monitoring. Properties of an indicator are *name* (the common name of the indicator), *identifier* (a unique value that allows to point to the indicator within the KPIOnto and PIKR), *acronym* (the acronym by which the indicator is known within the community), *definition* (a plain text giving a detailed description of the indicator and its use) and *unit of measurement* (the symbol and the description of the unit chosen for the indicator, making reference to an external ontology, namely the Measurement Units Ontology[4]).

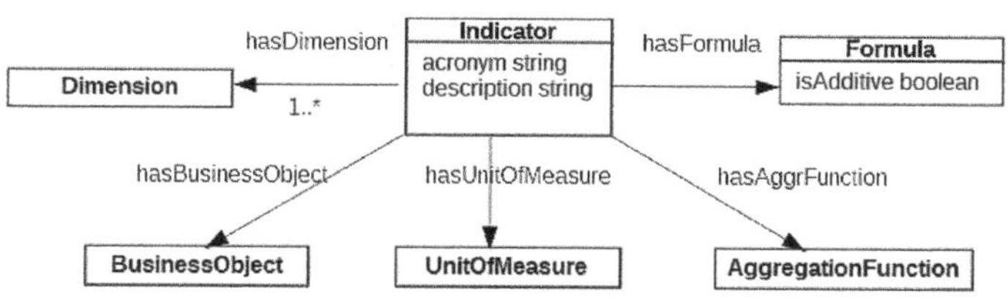

Figure 11.2. *The KPIOnto schema*

The class is further split into subclasses according to the classification provided by VRM. Five main subclasses have been individuated, dividing indicators in Corporate, Customer, Operation and Product performances. Part of the taxonomy is shown in Figure 11.3. The taxonomy of indicators can be subject to changes and extensions; for example, the possibility of defining multiple classifications on the basis of different models can be investigated.

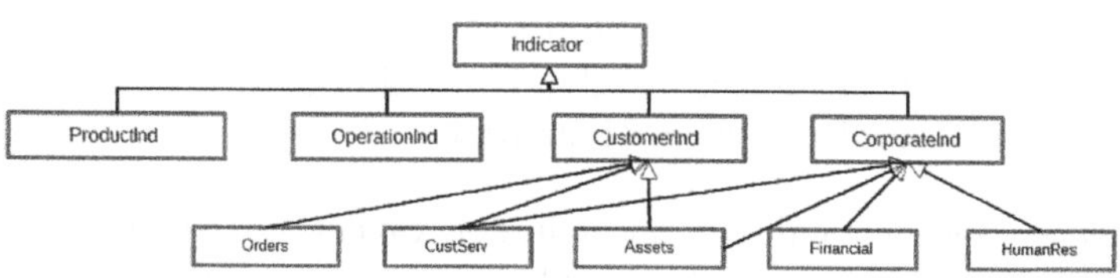

Figure 11.3. *The VRM taxonomy of indicators*

4 Measurement Units Ontology: http://idi.fundacionctic.org/muo/muo-vocab.html.

BusinessObject: following the VRM definition, a priority dimension describes the goal of optimization for which the indicator is used. VRM identifies the following instances of the Priority class: *reliability* (delivery of correct products to the correct market and customers on time), *velocity* (the cycle time to deliver a product or service to the customer), *adaptability* (response to market changes to gain or maintain competitive advantage), *cost* (associated with operating a value chain), *asset* (the effectiveness of an organization in managing assets of the value chain to support market), *innovation* (the ability to strategically leverage internal and external sources of ideas and introduce them to market through multiple paths) and *customer* (development of positive collaborative customer relationships).

Dimension is the coordinate/perspective to which the metric refers. For instance, it is useful to analyze deliveries along the delivery date (time dimension) or the delivered product (product dimension). Following the multidimensional model, a dimension is here structured into a hierarchy of *level* subclasses, where each level represents a different way of grouping elements of the dimension [KIM 02]. The dimension's hierarchy is represented by a *part-of* relationship among classes. Hence, for instance, the Organization dimension has a level Person that is part of a Team or of a Department, which in turn are part of an Enterprise. This idea is shown in Figure 11.4. Instances of a level are known as *members*, e.g. the company "ACME", the weeks "3rd-2012" and "42nd-2011". This ontological description of dimensions enables usual analysis operations over indicators, namely *roll-up* and *drill-down* (i.e. the moving from one level to its upper or lower level), that have a correspondence, respectively, with the part-of and its inverse relationship. Due to the existence of possible branched dimensions (see Figure 11.4), Person<Department<Enterprise and Person<ProjectTeam<Enterprise, the derivation of this path is not straightforward. In our proposal, the drill-down of a member *M* at level L_1 is the set of instances of the lower level (L_2), which belongs to the class described by the following expression: $\exists \text{partof}.\{M\} \sqcap L_2$. Referring to the example of Figure 11.4, $\exists \text{partof}.\{\text{ACME}\}=\{\text{R\&D}, \text{A\_Team}\}$, so $\exists \text{partof}.\{\text{ACME}\} \sqcap \text{Department}=\{\text{R\&D}\}$ is the drill-down of ACME along the hierarchy Person<Department<Enterprise.

Formula: the semantic of an indicator cannot be completely given without the representation of its formula, which is the way the indicator is computed. Each formula is characterized by the related indicator, the aggregation function, its graphical representation, its semantics (i.e. the mathematical meaning) and the references to its components, which are, in turn, formulas.

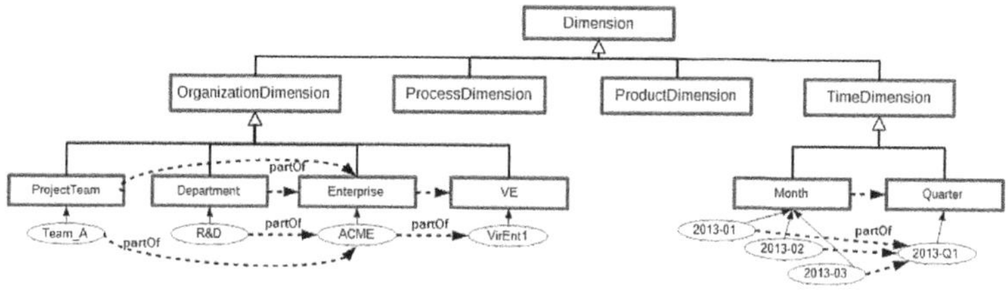

Figure 11.4. *An example of dimension hierarchy*

The formal representation and manipulation of the structure of a formula is essential in VEs to check inconsistencies among independent indicator definitions, reconcile indicators values coming from different sources and provide the necessary flexibility to indicators management. To this end, as explained in section 11.5, formula manipulation will be performed in logic programming. Furthermore, we resort to existing mathematical standards, namely MathML[5], for describing mathematical notations, capturing both its structure and content, and OpenMath[6] to represent operators semantics. Hereafter, we report an example of formula related to Idea Yield indicator.

```
Formula: kpi:Formula:name:IY1
kpi:Formula:presentation:
        <annotation-xml encoding="MathML-Presentation">
                <mi> IdeaYield </mi>
                <mo>&equals; </mo>
                <mfrac>
                        <mi> ideas_accepted </mi>
                        <mi> ideas_created </mi>
                </mfrac>
        </annotation-xml>
    kpi:Formula:content:
        <annotation-xml encoding="MathML-Content">
                <apply><eq/>
                        <ci> IY1 </ci>
                        <apply><times/>
                        <apply><divide/>
                        <ci> IAcc </ci><ci> ICre </ci></apply>
                        <cn>100</cn></apply>
```

5 W3C Math Working Group, *Mathematical Markup Language (MathML) Version 2.0 (Second Edition)*, http://www.w3.org/Math/.
6 OpenMath Society, *The OpenMath Standard Ver. 2.0*, http://www.openmath.org/.

```
                    </apply>
            </annotation-xml>
            <annotation-xml encoding="OpenMath">
                    <OMA
xmlns="http://www.openmath.org/OpenMath">
                            <OMS cd="arith1" name="times"/>
                            <OMS cd="arith1" name="divide"/>
                            <OMS cd="relation1" name="eq"/>
            </annotation-xml>
```

First, the "MathML-Presentation" part is introduced to represent the way the formula should be rendered. The "MathML-Content" section describes the semantics of the formula, which is the scope of operators. The formal definition for operators is given by referring to OpenMath Content Dictionaries (cd), which are collection of mathematical symbols and their definition. Finally, in the MathML-Content section, symbols "IAcc" and "ICre" are implicit references to corresponding indicators. On the basis of references, we are able to build up a graph of dependencies among formulas (and hence among related indicators). In this graph, a node without direct successors represents an *atomic indicator*, which is obtained as direct aggregation of transactional data.

11.5. Semantic services

Reasoning about KPIs is mainly based on the capability to manipulate formulas according to strict mathematical axioms, such as commutativity, associativity and distributivity of binary operators, and properties of equality needed to solve equations. Therefore, in order to define KPI reasoning functionalities, we need to represent both KPIOnto formulas and mathematical axioms in a common logic-based layer. To this end, we refer to the first-order logic and define the functionalities in logic programming (LP); in particular, we use Prolog and we refer to XSB[7], a logic programming and very efficient deductive database system, as the reasoning engine. For what concerns facts in the LP knowledge base, KPIOnto formulas are translated from MathML-Content to LP facts by using the predicates:

- `formula(Y, Expression, K)`, where *Y* is the indicator, *Expression* is its formula and *K* is a constant identifying whether *Y* is a leaf indicator or not;

7 XSB: http://xsb.sourceforge.net/.

- `equation(Equation, X)`, which represents a generic equation in the variable *X*.

For concerning predicates capable of manipulating formulas, we extended the PRolog Equation Solving System (PRESS) [STE 89] for the representation of mathematical axioms in LP predicates, as described in the next section, while in section 11.5.2 we introduce predicates to check the consistency of the repository.

11.5.1. *Formula manipulation*

Formula manipulation consists of deriving relations among indicators and rewriting formulas. The main types of predicates used in this work, on which the predicates introduced in section 11.5.2 are based, are useful for the simplification and for the resolution of equations. The former are exploited both to compact and improve the rendering of a formula by individuating and collecting equivalent terms (e.g. a*b and b*a, or a*a and a^{2}).

The main predicates of this kind are *simplify_term*, which is used to simplify equations, and *simplify_solution*, which is capable of rewriting the final solution in a more understandable form. The second type of predicates enables the symbolic resolution of equations by applying mathematical properties (e.g. commutativity, distributivity and factorization). The resolution of an "Equation" in a given variable "X" is realized through the predicate `solve_equation(Equation,X, X=solution)`, which is based on `expand_equation(Equation,ExpandedEquation)`, that rewrites the *Equation* by recursively replacing its operands by their formulas. The process ends when all operands of *ExpandedEquation* are atomic indicators (leaf nodes). In total, in order to perform the formula manipulation functionalities, more than 900 predicates are used in the theory.

11.5.2. *Consistency check*

In this chapter, we defined an indicator to be consistent with respect to the ontology if its definition is unique in the ontology, its formula is equivalent to a formula already in KPIOnto and its formula does not contradict any other already defined formulas. Given that the ontology must be guaranteed to be always consistent, the consistency is checked whenever an indicator is introduced, edited or deleted. In this way, enterprises always have an ontology available containing valid and usable information. These characteristics are of utmost importance in the domain, given that indicators are used to make decisions, often strategic. In the following, we discuss predicates that we introduced to check these conditions:

– `identical(Equation,X,X=Solution)`: this checks whether a fact formula(X,Expression,_) has been already added into the ontology;

– `equivalence(Equation,X,N)`: this is useful to individuate and manage duplications. The predicate finds all formulas for variable *X* that are equivalent to Equation by exploiting the *solve_equation* predicate;

– `incoherence(Equation,X,Y=S)`: this is used to verify that the modification or deletion of a formula does not contradict the ontology, i.e. it is coherent with the other previously defined formulas. Given the formulas in the ontology that are rewritable as functions of X, each element that is different from the Equation (after the appropriate expansion) leads to incoherence.

To provide some examples, let us consider the following indicators defined at VE level to measure costs and time spent for training of employees:

– PersonnelCosts=NumHours*HourlyCost*(Overhead+1);

– PersonnelTrainingCosts=HourlyCost*PersonnelTrainingTime;

– TeachCosts=PersonnelTrainingTime*HourRate;

– InvestmentInEmplDevelopment=PersonnelTrainingCosts+TeachCost.

Let `TotCostsEmpTrain` be a new indicator that an enterprise asks to add to the repository, with formula TeachCosts + PersonnelTrainingTime * HourlyCost. The three consistency predicates are executed in order. As no existing identical formula can be retrieved, the equivalence predicate is evaluated: by using the formula about "PersonnelTrainingCosts" and after some formula manipulation, the reasoner discovers that "TotCostsEmpTrain" is mathematically equivalent to "InvestmentInEmplDevelopment". Then, let us consider a user that requires to update "PersonnelTrainingTime", making a mistake when typing the formula (inserting a sum instead of a division): PersonnelTrainingTime = PersonnelTrainingCosts+HourlyCost. In this case, the new formula is neither identical nor equivalent to any formula, but it is incoherent with the definition of "PersonnelTrainingCost" together with every other higher-level indicators depending on it, i.e. "InvestmentInEmplDevelopment".

11.6. Semantic data handler

In this section, we present the SDH, the PIKR component enabling KPIs evaluation in the BIVEE scenario.

11.6.1. *Query management*

The SDH allows other BIVEE modules to execute OLAP-like queries, which are suited both for monitoring the evolution of processes and for analysis, enabling the user to unpack the data-cube to understand how a specific value has been generated. Following the underlying model, a multidimensional query is posed by specifying the following information:

– the name of the requested indicator (e.g. Costs_DeliverPhase);

– for each compatible dimension, the specific level on which to aggregate (e.g. Cat for ProductDimension and Month for TimeDimension);

– an optional set of members on which to filter (e.g. {Armchair} for "Cat" and {2013-01,...,2013-12} for "Month");

– the name of the requested VE and the names of the requested enterprises (e.g. ACME1 and ACME2);

– an optional flag specifying whether to obtain results for each enterprise separately or to aggregate them.

The SDH will return a table where each row represents the value of the indicator at the requested members (in previous example, the pairs <2013-01, Armchair> and <2013-02, Armchair>). It is worth noting that enterprises providing values are part of a particular dimension of analysis, namely the source, describing both enterprises and the VE to which the enterprise belongs. Since the BIVEE platform could manage data of various VEs, we explicitly need the information about the source, while other dimensions could be omitted in the query, requesting *de facto* to aggregate the whole cube in a unique value.

The theoretical framework of query resolution is that of query rewriting based on views (see, e.g. [HAL 01] in the context of distributed database systems, and [COH 06] and [GOL 12] for data warehouse integration). Views are dynamically defined by reasoning mechanisms over formulas and dimensions as defined in the KPIOnto. This allows us to reformulate the original query into a set of queries over local cubes where the actual levels and indicators provided by the enterprise are used. The methodology follows a two-stage approach, where at the first stage a completeness check is performed in order to ensure the quality of the result and, as a side effect, to determine which are the missing values. For these values, alternative definitions in terms of queries are then provided by exploiting two complementary expansion rules, and completeness is recursively checked. This produces a set of view definitions that combines together to generate the final nested query to the RDH. The two rules are:

1) indicator expansion: a new formula is inferred so that the value of the requested indicator at a given level can be calculated by other indicator values at the same level;

2) roll-up expansion: this rule allows to express the value of the requested indicator as an aggregation of values of the same indicator at lower level. We can further distinguish two variants of the rule, depending on the nature of the aggregation function: (1) for distributive functions, next lower level can be used, and (2) for algebraic functions, the lowest level expansion must be used.

The range of applicability of these rules depends on the type of aggregation function. Although with distributive aggregation functions (e.g. SUM) it is possible to aggregate a value starting from values pre-aggregated at its lower level (e.g. value for a quarter can be computed by summing values for months), in the general case, however, the correct value of an indicator at a given level can be only obtained by calculating the formula at the lowest level of granularity, and then applying the aggregation on the resulting values. Completeness check in traditional data warehouse systems is typically not supported, since data are simply produced by aggregating over the available values. This generates inexact results as long as managers are unaware of this fact. For example, obtaining a total cost of 100 at the first semester of 2013 and 500 at the second semester of 2013 can be interpreted erroneously as a sudden increase of costs, while it turns out that the total cost for the first semester has been calculated by summing over the first two months only if the others are not available. Since the probability of missing values is high in a VE environment, we believe that awareness of the completeness degree is a fundamental feature of a supporting environment.

11.6.2. *Architecture of the semantic data handler*

At a high level, the basic modules of the SDH are (1) *Correctness Checker*, (2) *Completeness Checker* and (3) *Query Execution*. As input, the SDH takes multidimensional queries from the MCR and VIF components.

The *Correctness Checker* module verifies the syntactic and semantic correctness of a query. The *Completeness Checker* module implements the completeness mechanism described in the previous section. These two modules interact with the PIKR Service Library for retrieving information from the PIKR ontologies and in particular the KPIOnto. Finally, the *Query Execution* module manages the actual calculation of the query by interacting with the RDH, through PIKR Connector. The interaction between modules is depicted in Figure 11.5.

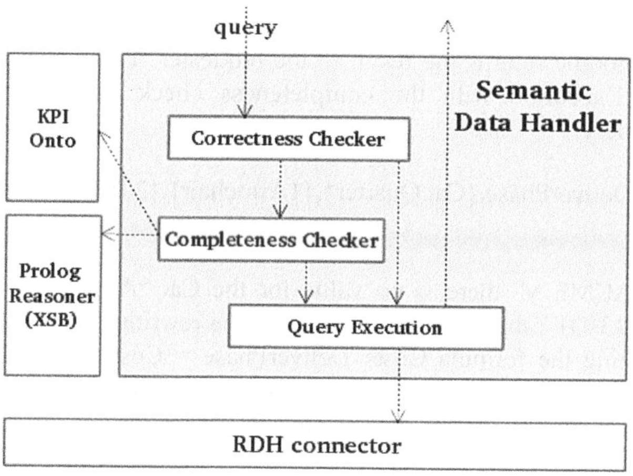

Figure 11.5. *Information flow in semantic data handler*

As for the Correctness Checker, the first step is performed by the *syntactic checker*, which is devoted to verify the syntactic correctness of a given multidimensional query by checking whether it is well formed, i.e. whether all required elements are provided and each element in the query is composed of terms in the KPIOnto. The verification of the semantic correctness of the query is the task of the *semantic checker* module, whose main goal is to guarantee that all specified terms are actually valid ontological concepts, and every specified member belongs to the correct level. Both syntactic and semantic checkers produce an error message for each of the above checks that fails, returning also the reason of the fail. All error messages are suitably packaged by the *Error Manager* module, and then returned to the caller of the Correctness Checker module.

After the verification of correctness of the multidimensional query, the SDH can execute the query in two ways, depending on the user request: as a "classical" query or as a "complete" query. In the former case, the Query Execution module is directly called. In the latter case, the query is first processed by the Completeness Checker module that checks whether it is complete and produces a new, more complex query that is passed to the Query Execution module.

The Query Execution module receives the original multidimensional query, which is first translated into SQL over the schema of the RDH. Then, the module interacts with the RDH in order to manage the query execution and, if necessary, properly processing the result set. Finally, the Query Execution module returns the result to the requester. To give an example of the Query Execution with the completeness check, let us consider the following query:

{Costs_DeliverPhase,{Cat,Quarter},{{Armchair},{2013-Q1}},{ACME_x, ACME_y}

If for "ACME_y" there is no value for the Cat "Armchair" and for the Quarter "2013-Q1", the system tries to apply the rewriting rules. Through rule 1 by exploiting the formula Costs_DeliverPhase = Costs_Ship+Costs_Pack+ Costs_OrPre+Costs_Deliv, a new query is automatically generated for each dependency of the KPI. In this case, the new queries are:

<Costs_Ship, {Cat,Quarter}, {{Armchair},{2013-Q1}},{ACME_y}>
<Costs_Pack,{Cat,Quarter},{{Armchair},{2013-Q1}},{ACME_y}>
<Costs_OrPre, {Cat,Quarter}, {{Armchair},{2013-Q1}},{ACME_y}>
<Costs_Deliv, {Cat,Quarter}, {{Armchair},{2013-Q1}},{ACME_y}>

If such queries are satisfiable, then the four values are retrieved and aggregated by using the formula. The result is the value for Costs_DeliverPhase requested by the original query. An alternative solution, according to rule 2, is to drill-down member 2013-Q1 or member Armchair to their lower members[8]. This implies, for every member, to create a new query and (if and once solved) to compute the required value from those available by means of the aggregation function of the indicator (if any). In the example, if the system drills-down 2013-Q1 to its components 2013-01, 2013-02 and 2013-03, and if the aggregator for the KPI is "Sum"", then it is

8 Relations among members that are used in this case are the part-of introduced in section 11.4.2.

possible to calculate the value as a summation of the results of the following queries:

<Costs_DeliverPhase, {Cat,Month}, {{Armchair},{2013-01}},{ACME_y}>
<Costs_DeliverPhase, {Cat,Month}, {{Armchair},{2013-02}},{ACME_y}>
<Costs_DeliverPhase, {Cat,Month}, {{Armchair},{2013-03}},{ACME_y}>

By executing the completeness check it is possible to obtain a larger number of results with respect to the classical query execution. However, due to its computational complexity, it is noteworthy that the running time of the completeness check procedure is related to the size of KPIOnto and the Data Storage. We refer the interested readers to [DIA 14a] for a detailed discussion on the indicator drill-down rule.

11.7. User applications

To support end-users in the management of KPIOnto and the extraction of KPI data for performance monitoring and comparison, we developed two web applications named KPIOnto Editor and KPIExplorer, described in the next sections.

11.7.1. *KPIOnto Editor*

KPIOnto Editor is an ontology editor which provides users with a graphical interface devoted to search for indicators, formulas and dimension's members previously stored in KPIOnto, as well as to create, modify and delete indicators and members, according to user privileges.

Introduction of a new KPI: as shown in Figure 11.6, the form allows us to insert both the descriptive and the structural definition of a KPI, limiting possible syntactic user errors, such as mistyped dimension's name or a formula referring to undefined KPIs. The user composes the formula by using a simple equation editor provided with a tool for searching the KPIs to include in the formula. After pressing on "Create", the formula is checked by using consistency predicates described in section 11.5.2.

Update of an existing KPI: this functionality is implemented in a way similar to the previous one. The formula of the KPI to update is temporarily removed from the ontology, and then the procedure for introducing a new KPI is executed. If the updated formula is consistent with respect to ontology, then the update is confirmed; otherwise, the old formula is restored. The interface implementing this functionality is the same as that presented in Figure 11.6, but the form is already filled with values from the ontology.

Deletion of a KPI: a KPI can be removed only if it does not make the ontology inconsistent. Hence, either the KPI is not part of any other formulas, or for each formula in which the KPI appears we can infer at least a new formula that does not contain KPI. Since all inferable formulas of an indicator are equivalent, we replace formulas containing the KPI by one of the inferred formulas not containing the KPI. Then, the indicator is removed from the ontology. The procedure is transparent to the user and does not require a specific interface. In the case the KPI cannot be deleted, the system returns formulas that cannot be rewritten.

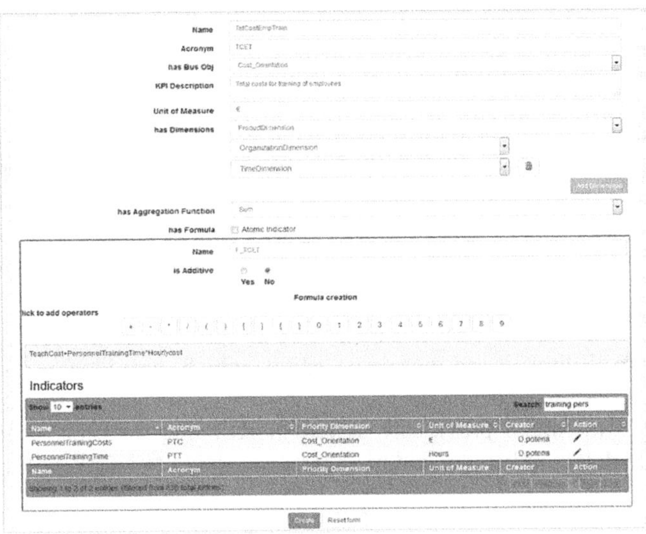

Figure 11.6. *KPIOnto Editor, introduction of a new KPI*

Other functionalities: the home page of the editor shows all indicators, formulas and members available in KPIOnto, together with a small set of properties; for indicators: the acronym, the priority dimension, the unit of measure and the creator; for formulas: name, a Boolean value indicating whether the formula is additive or not, the indicator for which the formula is used and the creator; for members: the name, the level, the dimension and the creator. By clicking on a specific indicator, formula or member, corresponding details are dynamically retrieved from the ontology and shown in a panel (see Figure 11.7) from which it can be modified or deleted. As for members, together with descriptive information, a graphical representation of the inferred part-of hierarchy is provided. The system also supports users when editing members, for instance by preventing them from defining wrong part-of hierarchies, which are not compliant with all the relations among members and levels that have been previously defined in the ontology.

11.7.2. *KPIExplorer*

KPIExplorer is a web-based tool that provides a semantically supported interface for OLAP analysis, following an ontology-based approach to access data. The tool supports users in the graphical composition of a multidimensional query, and allows its rewriting and execution on the data storage also exploiting dynamic, transparent and automatic calculation of KPI values by means of the SDH.

Through the user interface shown in Figure 11.8, the users graphically compose a multidimensional query to analyze KPIs for one or more enterprises. Once the required indicator within the KPI Selection search box is specified, the rest of the form is dynamically loaded, including the list of compatible dimensions. Dimension and their levels, as well as members defining filtering conditions, can be selected.

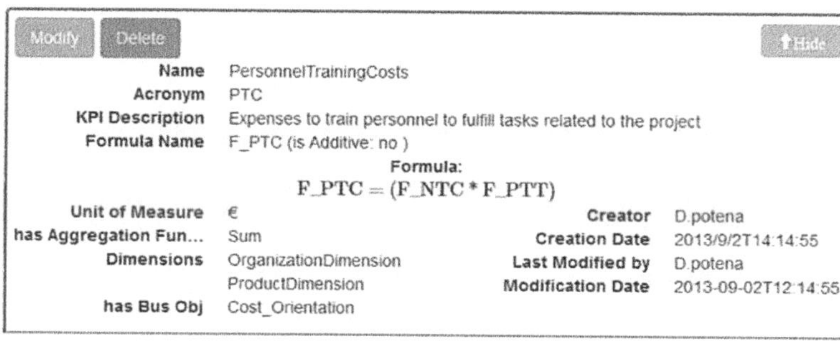

Figure 11.7. *KPIOnto Editor, details of a KPI*

In Figure 11.8, the user decides to analyze Costs_DeliverPhase KPI for enterprises ACME1 and ACME2, with respect to ProductDimension and TimeDimension. Concerning the former, the user selects "Cat" level and specifies "Armchair" as member on which to filter the results. As for the latter, the user specifies the first three months in 2013. Unselected dimensions are not part of the query. The Submit Query box allows us to run the query with two further features: by exploiting completeness check procedure and aggregating the query results.

Figure 11.9 shows the output of the submitted query. In this part of the interface, the "+" button besides each dimension's level allows us to enable classical dimension drill-down, whereas the "+" button beside the *Costs_DeliverPhase* KPI enables a novel drill-down functionality, namely *indicator drill-down* for expanding it into its components. By drilling-down the indicator, a new table is shown (as in Figure 11.10), where details about costs of shipping, packaging, costs of order preparation and costs of delivery forming the overall costs of the delivery phase are reported.

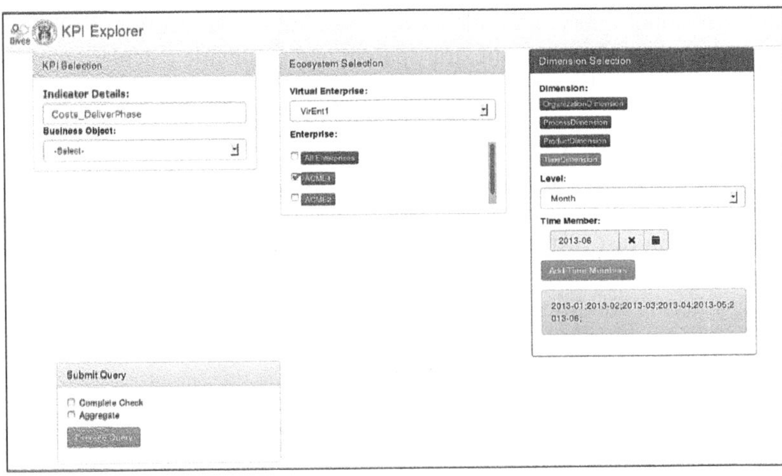

Figure 11.8. *KPIExplorer: composition of a query*

Cat	Month	Enterprise	Costs_DeliverPhase
Armchair	2013-01	ACME1	15.7
Armchair	2013-02	ACME1	15.1
Armchair	2013-03	ACME1	15.3
Armchair	2013-04	ACME1	14.8
Armchair	2013-05	ACME1	13.9
Armchair	2013-06	ACME1	14.3

Figure 11.9. *KPIExplorer: result table*

Indicator drill-down allows us to easily improve the user awareness about which specific subindicator has the major impact on the final cost. Since the data storage contains partially materialized indicators coming from possibly heterogeneous information systems, it is possible that values for some subindicators do not exist. This is the reason for empty cells in the figure.

Cat	Month	Enterprise	Costs_DeliverPhase				
				Costs_Ship	Costs_Pack	Costs_OrPre	Costs_Deliv
Armchair	2013-01	ACME1	15.7	6.23	2.15	3.1	4.22
Armchair	2013-02	ACME1	15.1	6.1		3.24	3.73
Armchair	2013-03	ACME1	15.3	6.03	2.23	3.09	3.95
Armchair	2013-04	ACME1	14.8				
Armchair	2013-05	ACME1	13.9	5.73	2.17	3.37	2.63
Armchair	2013-06	ACME1	14.3	5.5	2.19	3.13	

Figure 11.10. *KPIExplorer: drill-down of indicator costs delivering phase*

The lack of information can be at least partially solved by resorting to queries with complete check enabled. Figure 11.11 sheds light on this functionality for the drill-down of the *Cost_Pack* indicator. Interested readers can refer to [DIA 14a] and [DIA 14b], respectively, for further details on the indicator drill-down and KPIExplorer.

Cat	Month	Enterprise	Costs_DeliverPhase							
				Costs_Ship	Costs_Pack				Costs_OrPre	Costs_Deliv
						Materials_Pack	Others_Pack	LabourCosts_Pack		
Armchair	2013-01	ACME1	15.7	6.23	2.15		0.1	1.31	3.1	4.22
Armchair	2013-02	ACME1	15.1	6.1		0.69		1.23	3.24	3.73
Armchair	2013-03	ACME1	15.3	6.03	2.23	0.62	0.63	0.98	3.09	3.95
Armchair	2013-04	ACME1	14.8			0.68	0.33	1.14		
Armchair	2013-05	ACME1	13.9	5.73	2.17	0.71		1.2	3.37	2.63
Armchair	2013-06	ACME1	14.3	5.5	2.19	0.72	0.4	1.07	3.13	

a)

Cat	Month	Enterprise	Costs_DeliverPhase							
				Costs_Ship	Costs_Pack				Costs_OrPre	Costs_Deliv
						Materials_Pack	Others_Pack	LabourCosts_Pack		
Armchair	2013-01	ACME1	15.7	6.23	2.15	0.74	0.1	1.31	3.1	4.22
Armchair	2013-02	ACME1	15.1	6.1		0.69	0.11	1.23	3.24	3.73
Armchair	2013-03	ACME1	15.3	6.03	2.23	0.62	0.63	0.98	3.09	3.95
Armchair	2013-04	ACME1	14.8			0.68	0.33	1.14		
Armchair	2013-05	ACME1	13.9	5.73	2.17	0.71	0.26	1.2	3.37	2.63
Armchair	2013-06	ACME1	14.3	5.5	2.19	0.72	0.4	1.07	3.13	

b)

Figure 11.11. *KPIExplorer: drill-down of Costs Pack a) without and b) with completeness check*

11.8. Conclusion

In this chapter, we discussed the design and implementation of semantics-based interoperability services aimed to enable performance monitoring in a VE environment. The proposed solution is mainly based on: (1) a KPI reference ontology (KPIOnto) to overcome heterogeneities among data for evaluating performance indicators, (2) a set of logic-based reasoning functionalities capable of manipulating KPI formulas and (3) the SDH, addressing other heterogeneities (e.g. structure and granularity). SDH provides a rewriting mechanism that, starting from a query based on the KPIOnto format, produces a query to retrieve actual data from the data storage. In this way, the approach enables automatic, dynamic and transparent aggregation of KPIs at any level, and computation of KPI values through their mathematical formulas, by exploiting information in the ontology.

Although only indicators with the analytic expressions can be currently represented, this does not limit the model applicability since other kinds of indicators (e.g. qualitative indicators) can be introduced as atomic. We plan to study extensions of the theory toward more complex expression manipulation. Other extensions regard the representation of relational algebra expressions as indicator formulas, and of different aggregation functions for different indicator dimensions. Future work will also include comprehensive tests aimed to determine the effectiveness and efficiency of the completeness check of the SDH. An early experimentation is available in [DIA 14a].

11.9. Bibliography

[ADA 06] ADAMS R., BESSANT J., PHELPS R., "Innovation management measurement: a review", *International Journal of Management Reviews*, vol. 8, no. 1, pp. 21–47, 2006.

[COH 06] COHEN S., NUTT W., SAGIV Y., "Rewriting queries with arbitrary aggregation functions using views", *ACM Transactions on Database Systems*, vol. 31, no. 2, pp. 672–715, 2006.

[CRU 05] CRUZ I.F., XIAO H., "The role of ontologies in data integration", *Journal of Engineering Intelligent Systems*, vol. 13, pp. 245–252, 2005.

[DEL 10] DEL-RÌO-ORTEGA A., RESINAS M., RUIZ-CORTES A., "Defining process performance indicators: an ontological approach", *Proceedings of the 2010 International Conference on the Move to Meaningful Internet Systems – Part I (OTM '10)*, Berlin, Heidelberg, Springer-Verlag, pp. 555–572, 2010.

[DIA 14a] DIAMANTINI C., POTENA D., STORTI E., "Extending drill-down through semantic reasoning on indicator formulas", in BELLATRECHE L., MOHANIA M. (eds), *DaWaK 2014,* Lecture Notes in Computer Science, Springer, vol. 8646, pp. 57–68, 2014.

[DIA 14b] DIAMANTINI C., POTENA D., STORTI E. *et al.*, "An ontology-based data exploration tool for key performance indicators", in MEERSMAN R. *et al.* (eds), *On the Move to Meaningful Internet Systems: OTM 2014 Conferences*, Lecture Notes in Computer Science, vol. 8841, Springer, pp. 727–744, 2014.

[GOL 12] GOLFARELLI M., MANDREOLI F., PENZO W. *et al.*, "OLAP query reformulation in peer-to-peer data warehousing", *Information Systems*, vol. 37, no. 5, pp. 393–411, 2012.

[HAL 01] HALEVY A., "Answering queries using views: a survey", *VLDB Journal*, vol. 10, no. 4, pp. 270–294, 2001.

[HOR 14] HORKOFF J., BARONE D., JIANG L. *et al.*, "Strategic business modeling: representation and reasoning", *Software & Systems Modeling*, vol. 13, no. 3, pp. 1015–1041, 2014.

[HUA 07] HUANG S.M., CHOU T.H., SENG J.L., "Data warehouse enhancement: a semantic cube model approach", *Information Sciences*, vol. 177, pp. 2238–2254, 2007.

[KEH 09] KEHLENBECK M., BREITNER M., "Ontology-based exchange and immediate application of business calculation definitions for online analytical processing", in PEDERSEN T., MOHANIA M., TJOA A. (eds), *Data Warehousing and Knowledge Discovery*, LNCS, vol. 5691, Springer Berlin Heidelberg, pp. 298–311, 2009.

[KIM 02] KIMBALL R., ROSS M., *The Data Warehouse Toolkit: The Complete Guide to Dimensional Modeling*, 2nd ed., John Wiley & Sons, Inc., New York, 2002.

[LAK 03] LAKSHMANAN L.V.S., PEI J., ZHAO Y., "Efficacious data cube exploration by semantic summarization and compression", *Proceedings of the 29th International Conference on Very Large Data Bases*, vol. 29, pp. 1125–1128, 2003.

[LEN 02] LENZERINI M., "Data integration: a theoretical perspective", *Proceedings of the 21st ACM SIGMOD-SIGACT-SIGART Symposium on Principles of Database Systems (PODS)*, ACM, pp. 233–246, 2002.

[MAS 07] MASAYNA V., KORONIOS A., GAO J. *et al.*, "Data quality and KPIs: a link to be established", *Proceedings of the 2nd World Congress on Engineering Asset Management (EAM)*, Harrogate, United Kingdom, 2007.

[PED 09] PEDRINACI C., DOMINGUE J., "Ontology-based metrics computation for business process analysis", *Proceedings of the 4th International Workshop on Semantic Business Process Management*, pp. 43–50, 2009.

[POP 10] POPOVA V., SHARPANSKYKH A, "Modeling organizational performance indicators", *Information Systems*, vol. 35, pp. 505–527, 2010.

[PRI 03] PRIEBE T., PERNUL G., "Ontology-based integration of OLAP and information retrieval", *Proceedings of DEXA Workshops*, pp. 610–614, 2003.

[STE 89] STERLING L., BUNDY A., BYRD L. *et al.*, "Solving symbolic equations with press", *Journal of Symbolic Computation*, vol. 7, no. 1, pp. 71–84, January 1989.

[TEN 01] TENNANT G., *Six Sigma: SPC and TQM in Manufacturing and Services*, Gower Publishing, Ltd., 2001.

[XIE 07] XIE G., YANG Y., LIU S. *et al.*, "EIAW: towards a business-friendly data warehouse using semantic web technologies", *The Semantic Web*, Lecture Notes in Computer Science, vol. 4825, Springer Berlin Heidelberg, pp. 857–870, 2007.

12

Raw Data Connection Services and Tools

Data interoperability is one of the top concerns in virtual enterprise environments. One specific aspect of this problem is the cross-organization measurement of business performance. Key performance indicators (KPIs) can depend on heterogeneous data maintained in different information systems: a software infrastructure is required in order to collect these data and to present it to consumers, both humans and ICT systems, in a homogeneous way. This chapter reports on BIVEE's raw data connection services and tools (RDCSs), which address these requirements by leveraging advanced extract-transform-load (ETL) techniques and a common KPI reference ontology.

12.1. Introduction

Virtual enterprise environments are, by their very nature, backed by distributed information systems. The term "distributed" means, in this context, not necessarily "geographically sparse" but rather "heterogeneous". As a matter of fact, virtual enterprises are not designed as such from the ground up: they aggregate existing organizations, each of them contributing their own assets and information resources. Not surprisingly, building a virtual enterprise is also a matter of data integration.

Chapter written by Mauro ISAJA.

Information can be literally anything (e.g. knowledge, product lifecycle data, etc.), but here we will focus on one specific aspect of governance: performance monitoring – i.e. the continuous assessment of how business processes are performing. To undertake performance monitoring, you first need to collect measurements from real-world activities, and then to transform these *raw data* into some meaningful *key performance indicator* (KPI) set. This is true for any enterprise, and virtual enterprises are no exception. However, here is where the heterogeneous nature of information gets in the way.

At the lowest level, performance-related information is stored in different systems in different ways. Measurements of comparable phenomena with similar semantics may differ in representation (structure and format), aggregation level and/or in the *dimensions* they refer to. However, the very nature of the virtual enterprise mandates a *holistic* view on performance, which means that heterogeneous data should be reconciled.

Raw data connection services and tools (RDCS in brief) are a composite subsystem of the BIVEE platform which addresses the heterogeneity problem of information at the *raw data* level. This means that *semantic lifting* and *normalization* techniques are applied to the original data sets, resulting in data that can be accessed from a centralized endpoint and navigated according to the standard model defined by the virtual enterprise. The process is minimally invasive with respect to the legacy systems involved, and is implemented by means of *IT* tools such as raw data management, semantic annotation and the extract-transform-load (ETL) development environment, which are detailed in the following sections.

12.2. Raw data management

The raw data management (*RDM*) is the module of the BIVEE platform which is responsible for importing and storing raw KPI data. RDM is composed of three layers: *data storage (RDM-DS), public API (RDM-WS) and front-end (RDM-FE)*. The relationship of RDM layers with other components of the BIVEE platform is shown in Figure 12.1.

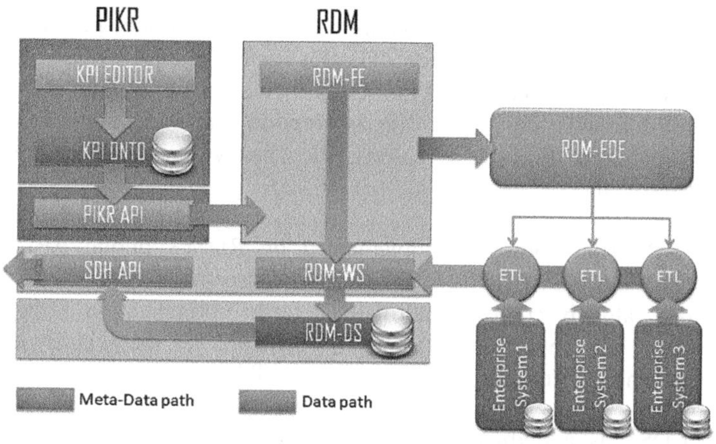

Figure 12.1. *Architecture of the RDM subsystem*

12.2.1. *Data storage*

Data storage – RDM-DS in brief – is where historical KPI data are kept, in *normalized* form; i.e. in a form that is highly compatible with the logical structure modeled by the reference ontology, *KPIOnto* [TAG 12]. The main value of data normalization is that application queries, expressed in KPIOnto's terms, can be translated into equivalent RDM-DS queries which will perform very fast (also see section 12.2.2).

Obviously, data normalization has a price: the transformation of data coming from different sources is a complex and time-consuming task. The key point, however, is that in RDCS the overhead of normalization is placed offline: the toll is paid once when data are imported; after that, repeated access is nearly free.

It should be noted that this approach, which can be summarized as "heavy loading, light reading", delivers its value when two conditions are met:

1) Loading is done infrequently in comparison with read access: source data do not change too frequently, or the users do not need a near-real-time view on ever changing data.

2) A significant portion of the loaded data records are actually read: relatively few computational resources are spent to process items that are then left unused.

In the BIVEE scenario, KPI data are typically loaded once per day, as the time granularity of a performance indicator measurement is rarely, if ever, shorter than 24 h. Moreover, KPIs are, as a rule, carefully crafted business intelligence formulas, so it is a safe assumption that most of them will be actually monitored. For these reasons, the "heavy loading, light reading" approach fits well.

Raw KPI data loading is implemented by RDM's public API, which imports into the RDM-DS a serialized data stream. When new data are submitted, it already comes in normalized form: normalization – the heavy job – is done on the sender side, more specifically by the ETL procedures running on sender premises. The ETL process will be described in section 12.3.

RDM-DS has a dedicated subschema for each KPI, according to the definition of the indicator in KPIOnto. As discussed in [TAG 12], an indicator in KPIOnto is associated with its formula and with one or more instances of dimension, representing those coordinates along which the indicator can be computed. A dimension is structured as a hierarchy of levels, where each level represents a different way of grouping elements of the dimension (e.g. city, region and country are levels of a geographic dimension). The dimension's hierarchy is represented by a part-of relation among levels. For instance, the "organization" dimension defines a hierarchy of levels: enterprise, department, team and person. An individual belonging to the person level will define a part-of relationship with respect to another individual belonging to the team level, who in turn will have a part-of relationship with a given department, etc.

It is expected that KPI data to be imported will have the same structure as the database it should fit in: any mismatch is a blocking issue, so structural validation is implicit in the data loading process. The import logic has then two more tasks to accomplish: verifying data constraints and reconstructing the hierarchy of dimension levels.

Verifying data constraints means checking that all required fields in each record have a value, and that all provided values are valid. Each KPI defines some mandatory properties – i.e. the KPI name and its value – and some mandatory dimensional references – i.e. time and virtual enterprise. Validity is checked for dimensional references only, and is a twofold task. First, the provided values must be valid by themselves – i.e. they should match an instance of the corresponding dimension level in KPIOnto, as replicated in RDM-DS tables. Then, if a dimension is defined as a hierarchy of levels, only one of those levels should be specified in data, leaving the system in charge of deriving the missing information.

This last task – reconstructing the hierarchy of dimension levels – poses some real challenges and issues. The actual hierarchy is inferred from the *partOf* relationships defined in KPIOnto, also replicated in RDM-DS meta-data. However, *partOf* relationships might be used in such a way that a given item has more than one *parent* on a higher level: in this case, the import logic will not be able to identify the correct one, and the data loading process will fail.

It should be noted that the entire data loading process is driven by meta-data: the import logic has no predefined knowledge of KPIOnto and the database structures used to store KPIs. This loose coupling between code and data structures is a key concept that, as we will see, is applied pervasively in RDCS, and makes the entire system adaptable to changes in the KPIOnto without requiring software updates.

The storage of KPI data and meta-data is managed through a relational database management system (RDBMS), and implemented leveraging on the *star schema* pattern – that is a highly denormalized relational data structure which trades space efficiency in favor of better runtime performance of analytic SQL queries [KIM 02].

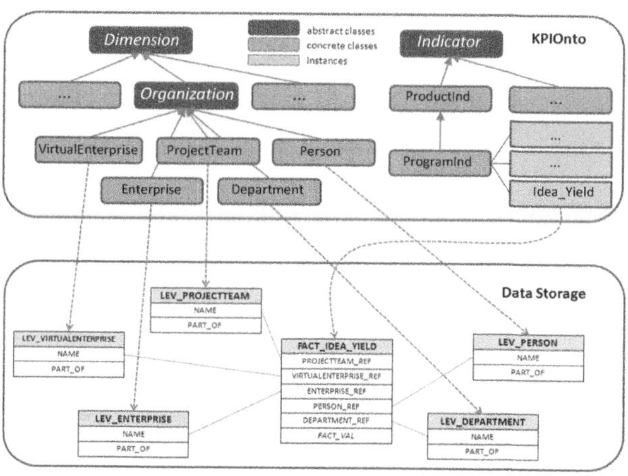

Figure 12.2. *Example of how ontology concepts are mapped to database schema*

The peculiarity of the RDM-DS implementation is that the database structures used to store KPIs are dynamically generated – and then populated with domain entities – using the reference ontology as a blueprint. Basically, KPIOnto dimensions/levels are translated into level tables and each indicator instance (i.e. a KPI definition) becomes a fact table by its own, connected with the relevant dimensions. The resulting schema is described in the data storage catalog and from there entities can be easily mapped back to KPIOnto concepts. Figure 12.3 illustrates this mapping strategy with an example.

Once the database schema is in place, level tables are finally populated with records, each corresponding to an instance of the corresponding KPIOnto class (see Figure 12.4).

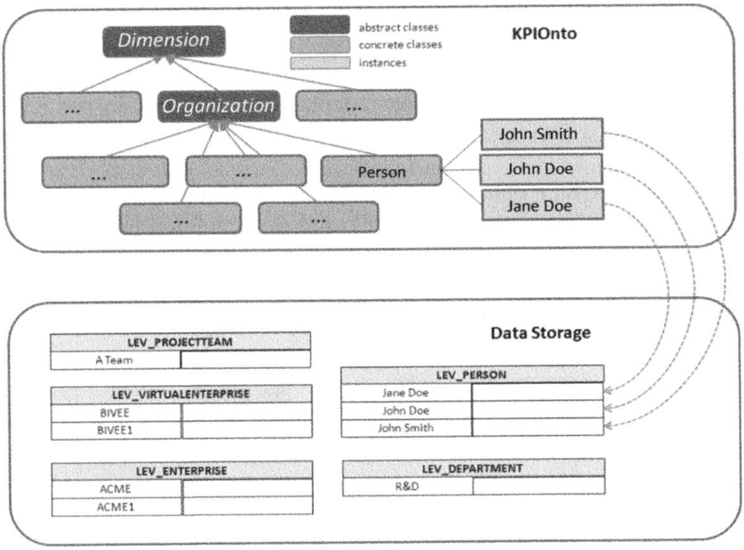

Figure 12.3. *Example of how ontology concepts are mapped to database records*

The above described approach has a shortcoming. In KPIOnto, dimensions can be partitioned in levels by subclassing, but hierarchical relationships are defined at the *instance* level by applying the *partOf* predicate. This enables a very flexible use of dimensions, such that it is possible to define hierarchies in KPIOnto (see example in Figure 12.4) that cannot be translated into an RDBMS multilevel dimension – at least not without complicating the database design up to the point of losing the advantages offered by the *plain* star schema (simple access path for aggregate queries).

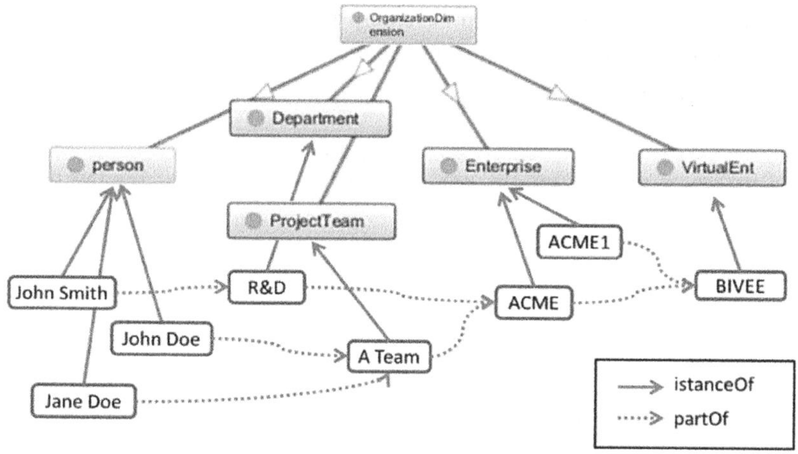

Figure 12.4. *Dimension hierarchies in the KPIOnto*

So, by design, all level tables in RDM-DS represent a single level of a KPIOnto dimension and store a flat list of names, which correspond to KPIOnto instances. In fact, the KPIOnto notion of the *partOf* relationship is indeed used in RDM-DS records (see example in Table 12.1), but just as additional meta-data; in practical terms, it cannot be directly exploited by SQL queries. However, this information is required to validate dimension-level hierarchies that are captured in loaded *facts*, as explained in Table 12.1.

LEV_PERSON	
NAME	**PART_OF**
John Doe	PROJECTTEAM:A Team
Jane Doe	PROJECTTEAM:A Team
John Smith	DEPARTMENT:R&D

Table 12.1. *Example of partOf relationship described in the data storage*

Capturing a dimension-level hierarchy in loaded facts means that KPI records implicitly express hierarchical relationships by pointing to dimension members placed on different levels. In the example given below, and shown in Table 12.2, four different facts are (indirectly) telling us that:

1) "John Doe" and "Jane Doe" are both members of the "A Team" project in the scope of the "ACME" enterprise, which in turn is member of the "BIVEE" virtual enterprise.

2) In the same virtual enterprise/enterprise scope, "John Smith" belongs to the "R&D" department and is not a member of any project.

3) "ACME1" is another member of the "BIVEE" virtual enterprise.

FACT_IDEA_YIELD							
FACT_ID	VIR...	ENT...	DEP...	PRO...	PER...	...	FACT_VAL
62935	BIVEE	ACME	null	A Team	Jane Doe	...	15.088
62936	BIVEE	ACME	R&D	null	John Smith	...	12.951
62937	BIVEE	ACME	null	A Team	John Doe	...	13.732
62938	BIVEE	ACME1	null	null	null	...	14.119

Table 12.2. *Multilevel dimension hierarchy as captured in facts*

Note that capturing relationship in facts is *not* the same as expressing it explicitly in a multilevel dimension table, as there is no guarantee that all hierarchies are fully described by fact records. So, in RDM-DS, we cannot use SQL queries to navigate multilevel dimensions by themselves – e.g. extract all the children of a given dimension member. However, we can still correctly aggregate KPI data and perform roll-up and drill-down operations.

12.2.2. *Public API*

The RDM module exposes its own public API: RDM web service (RDM-WS). This comes in two *flavors*: a HTTP/SOAP version and a HTTP/REST version. Both provide identical functionality, but the different protocols target different environments, allowing for the integration of a wider array of clients/technologies (e.g. Java-based on HTTP/SOAP and JavaScript on HTTP/REST).

RDM-WS enables BIVEE applications, the ETL jobs developed in the integrated development environment (see section 12.3), or even any (authorized) third-party application to move bulk KPI data to and from the RDM-DS storage. The following is a summary list of the available API calls. For technical details, please refer to the BIVEE environment's prototype documentation [ISA 14]:

– retrieve all KPI definitions;

– retrieve all the dimensions of a given KPI definition;

– retrieve all the members of a given dimension;

– retrieve all facts belonging to a given KPI definition;

– import facts belonging to a given KPI definition;

– delete a fact belonging to a given KPI definition.

Besides the direct access path provided by RDM-WS, typically KPI data stored in RDM-DS are served to BIVEE applications through the mediation of the *semantic data handler (SDH)*, which is a key component of the Production and Innovation Knowledge Repository (PIKR) [TAG 13]. SDH has its own notion of RDM-DS' relational database schema, as this is modeled from KPIOnto definitions using a well-known set of rules and naming conventions:queries expressed in KPIOnto terms are then automatically translated into SQL, and executed into RDM-DS in a very efficient way.

12.2.3. *Front-end*

Finally, RDM-FE is web-based front-end user interface (UI) for administrative purposes, and for the manual input of KPI facts into RDM-DS when no automated ETL job is available. It provides basic navigation and edit functionality as documented in Figures 12.5 and 12.6.

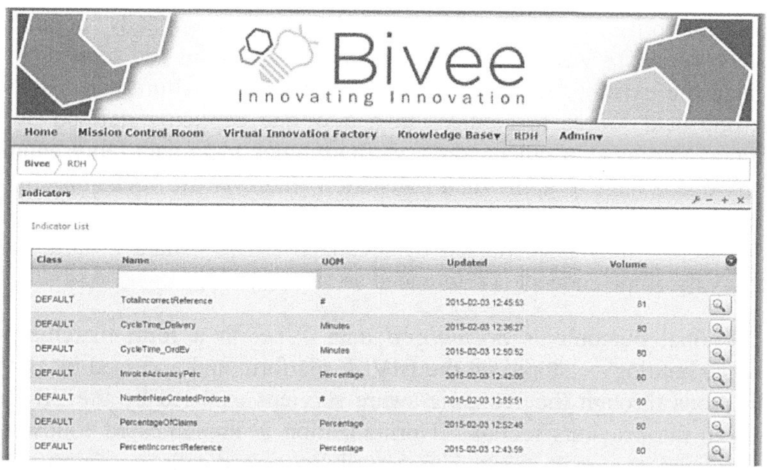

Figure 12.5. *RDM-FE: KPI list*

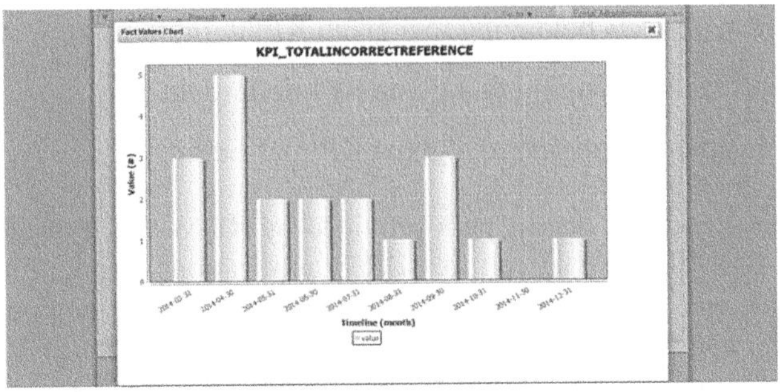

Figure 12.6. *RDM-FE: KPI fact history (dialog)*

12.3. Semantic annotation and ETL development environment

In general terms, *semantic annotation* is about enriching data (i.e. annotate/tag content items) with meta-data which map heterogeneous information to the structured knowledge of a domain, as formally defined by an ontology. Applying semantic annotation results in *semantic lifting*: unstructured content turned into semantic knowledge resources.

In the context of RDCS, this wide-scoped concept is much more focused. The requirement is to gather corporate-level operational and innovation process indicators (KPIs), extracted from different systems and domains, and transform/reshape them so they fit into a common model defined at the virtual enterprise level. This process is more appropriately defined as *semantic re-engineering of structured data*, as we do not add semantics: we just map the existing, which is implicit in the data structure and may vary from one data source to another, to a new one explicitly defined by the reference ontology. In other words, we harmonize meta-data representation. Thus, we also have to apply data translation wherever the vocabulary in use at the source differs from that on the ontology – e.g. when in source data the city of Rome, Italy is identified by localized string "Roma", while in the ontology the same concept is referenced as "Rome".

The final outcome is *normalized* data – i.e. in a form matching the reference ontology – stored on the BIVEE platform and accessed by BIVEE applications through the semantic-aware services exposed by the PIKR, as shown in the previous sections. Normalization is implemented following a decentralized approach: each data contributor – typically an enterprise *IT* service – designs, deploys and runs its own set of ETL procedures which extract local data, transform them in a normalized form and finally load them on the RDM module of the BIVEE platform.

To support this strategy, RDCS provides the *ETL development environment (RDM-EDE)*. This is a desktop application for the assisted development of *ETL jobs* capable of *feeding* RDM with normalized KPI data.

This decentralized approach provides some key advantages:

1) Each data contributor maintains in-house its own transformation rules (the semantic annotation layer).

2) Each data contributor runs the most computationally intensive operations on its own hardware, thus distributing the load across the entire BIVEE network.

RDM-EDE is a BIVEE extension of the Talend Open Studio for Data Integration[1] (*TOS*) open source integrated development environment (IDE). While this platform is targeted at the IT specialist, building ETL procedures does not require programming skills. The RDM-EDE desktop application is typically installed on the workstation of a corporate user.

In Figure 12.7, a snapshot of the TOS platform is shown. The three arrows superimposed on the screen snapshot point to the BIVEE extensions.

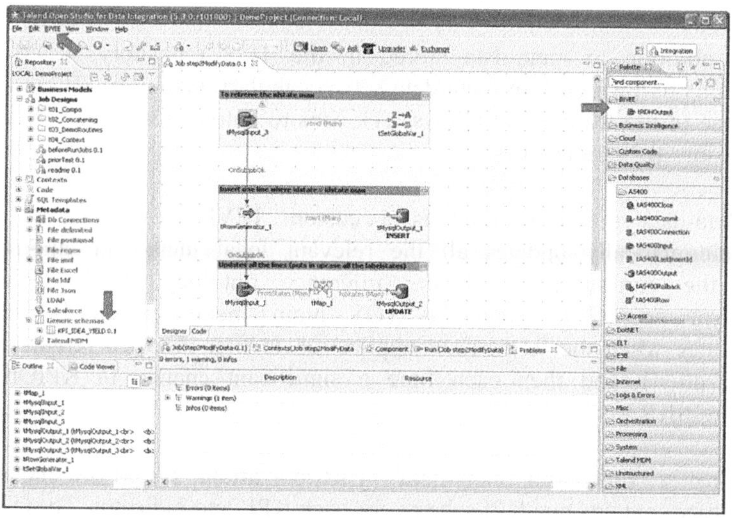

Figure 12.7. *The Talend Open Studio platform*

1 http://www.talend.com/products/data-integration.

BIVEE RDCS extensions to TOS are related to four aspects:

– *Meta-data synchronization*: an integrated utility is used to keep TOS meta-data in sync with KPIOnto.

– *Data translation*: a dedicated graphical user interface (GUI) functionality enables the user to set rules on how terms of the *local* vocabulary (the enterprise domain) are automatically translated by ETL jobs into their *global* equivalent, as defined by the reference ontology (KPIOnto).

– *ETL job building*: a custom job designer component connects the data output of a job to the data input of RDM.

– *ETL job* deployment: an integrated utility is used to transform *regular* ETL procedures, developed in the job designer, into BIVEE ETL jobs.

12.3.1. *Meta-data synchronization*

In RDM-EDE, TOS meta-data are used to describe KPIOnto indicators: each KPI defined in the reference ontology is represented by a specific, autogenerated generic schema item.

In addition to meta-data stored in the TOS repository, RDM-EDE maintains its own custom meta-data set: vocabulary mappings. These are at the base of the feature that allows ETL jobs to automatically translate local names into KPIOnto names.

Meta-data synchronization is a custom BIVEE extension to TOS that automatically updates all the relevant local meta-data, aligning it with the current state of the reference ontology. It is a utility program, accessible from the TOS main menu, which is run at least once when the RDM-EDE application is first installed (environment initialization), and then each time a significant change to KPIOnto is done in PIKR.

After a successful meta-data synchronization, the TOS meta-data repository contains one KPI schema for each KPIOnto indicator. These items are named after the original indicator, with the addition of a "KPI_" prefix (see Figure 12.8).

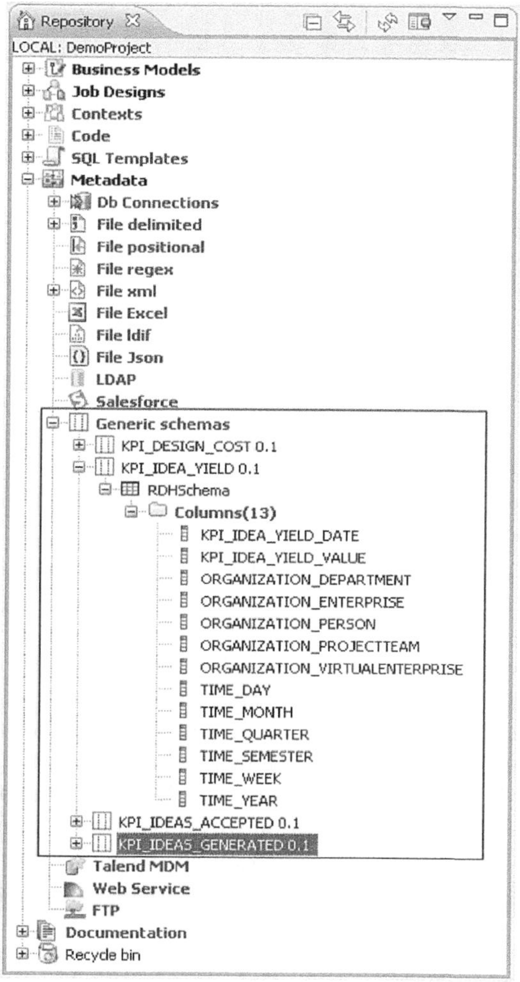

Figure 12.8. *Example of KPI schemas*

A KPI schema is automatically constructed by adding to a base set of invariant columns a number of extra columns derived from the dimensions supported by the corresponding KPIOnto indicator. More specifically, the schema will contain one column for each level of each supported dimension, and one numerical column for the measurement values.

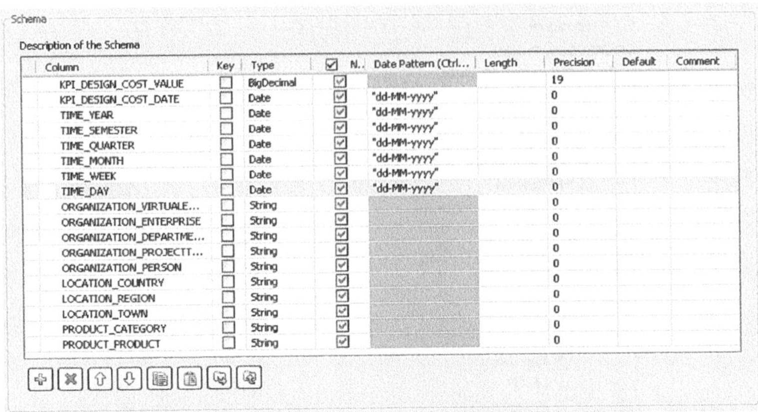

Schema

Description of the Schema

Column	Key	Type	N..	Date Pattern (Ctrl...	Length	Precision	Default	Comment
KPI_DESIGN_COST_VALUE		BigDecimal				19		
KPI_DESIGN_COST_DATE		Date		"dd-MM-yyyy"		0		
TIME_YEAR		Date		"dd-MM-yyyy"		0		
TIME_SEMESTER		Date		"dd-MM-yyyy"		0		
TIME_QUARTER		Date		"dd-MM-yyyy"		0		
TIME_MONTH		Date		"dd-MM-yyyy"		0		
TIME_WEEK		Date		"dd-MM-yyyy"		0		
TIME_DAY		Date		"dd-MM-yyyy"		0		
ORGANIZATION_VIRTUALE...		String				0		
ORGANIZATION_ENTERPRISE		String				0		
ORGANIZATION_DEPARTME...		String				0		
ORGANIZATION_PROJECTT...		String				0		
ORGANIZATION_PERSON		String				0		
LOCATION_COUNTRY		String				0		
LOCATION_REGION		String				0		
LOCATION_TOWN		String				0		
PRODUCT_CATEGORY		String				0		
PRODUCT_PRODUCT		String				0		

Figure 12.9. *Example of KPI schema in detail*

As with schema names, column names are determined by convention from the pattern <dimension_name>_<level_name>, except that for the value column, which follows the <kpi_name>_VALUE form (see Figure 12.9).

An important warning that must be taken into account by the user is that all meta-data synchronizations performed after the first one (the typical scenario of an ontology update) might invalidate existing jobs. For example, a new level might be added to a dimension, so that all KPIs using that dimension now have a new schema with an additional column – not handled by existing jobs. This is also the reason why this action is manual: RDCS has no automated mechanism for propagating changes from KPIOnto to RDM-EDE.

12.3.2. *Data translation*

As already mentioned in the introduction to the semantic annotation topic, data translation is the second pillar of semantic re-engineering of structured data, the first being schema mapping. The former is addressed by this section, while the latter will be discussed in the next section.

Data translation is based on vocabulary mappings, an important meta-data resource in RDM-EDE: local tables where dimension-level members of the local domain are matched to the corresponding KPIOnto entities. Vocabulary mappings are defined by the user in the project setup phase. Figure 12.10 shows when data normalization rules are defined and when they are applied.

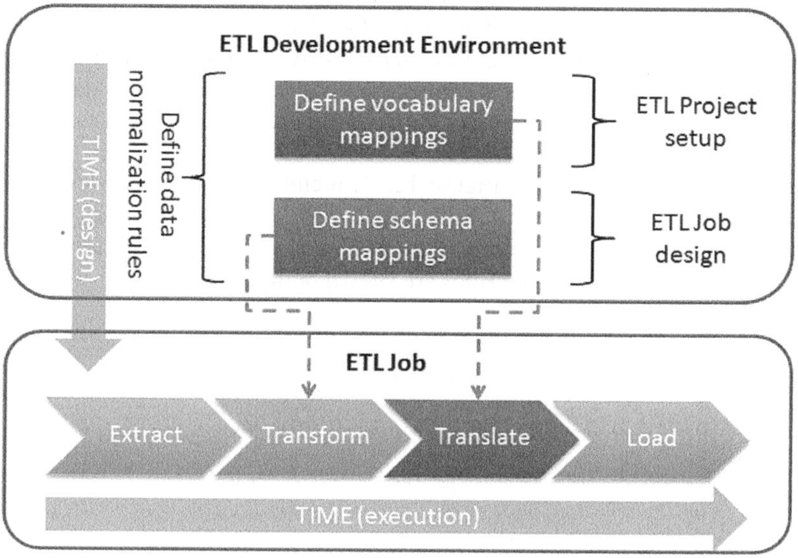

Figure 12.10. *Phases of data normalization*

Vocabulary mapping definitions are handled by a custom BIVEE extension, accessible from the TOS main menu. From the user point of view, this tool is a GUI where each dimension level in KPIOnto is a list item, bound to a dedicated mapping table. In order to edit a mapping table, which is initially empty, the user first selects the relevant dimension level from the list. The mapping tool then queries both the source system and PIKR, identifying all the names in a given local domain (determined by the user), which have no direct match and no mapping with any peer item in the dimension level (KPIOnto instances): the result is a list of unmatched names. The user goes through the list of unmatched names, and for each one selects its KPIOnto equivalent. Finally, the mapping table is saved locally as an XML resource.

In the use case described above, there are two notable points:

1) The user needs to explicitly map only those local names that actually differ from their global equivalent.

2) It is assumed that the mapping tool can extract all distinct names from a given domain of the source system. The tool provides means to define an extraction SQL query and DB connection details. So, it relies on the assumption that the source system is an SQL database which can be reached through a java database connectivity (JDBC) driver, which in turn is installed in RDM-EDE's environment. If this is not the case – i.e. local data come from a legacy system – local names must be imported manually.

In the example shown in Figure 12.11, members of the three-level local GEO_DIM dimension are matched to their KPIOnto counterparties, and the outcome is that only one member per level actually needs translation – the resulting vocabulary mapping rules are shown bottom right.

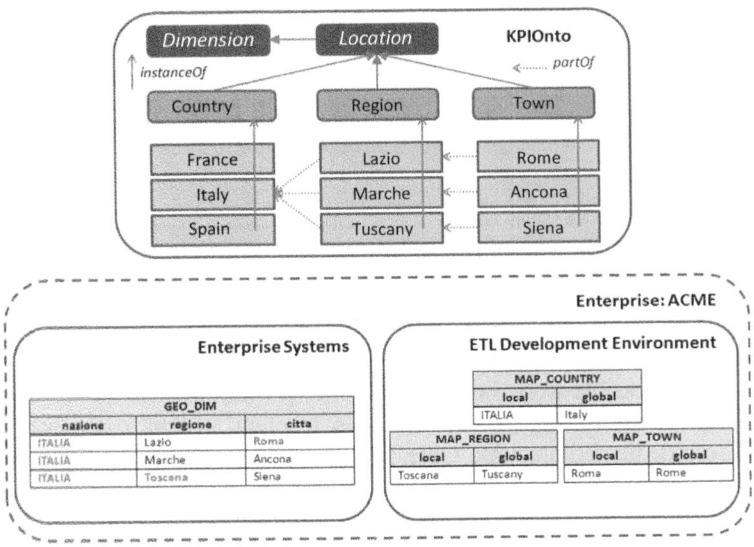

Figure 12.11. *Data translation example*

At runtime, when the ETL job is executed, normalization rules are applied to source data: schema mapping is always taken into account, while data translation happens only when actually required. In this phase, any irresolvable mismatch between source data and the reference ontology is a fatal error. In other words, vocabulary mappings defined at project setup must cover all cases represented in data, or data loading in RDM-DS will fail (see Figure 12.12).

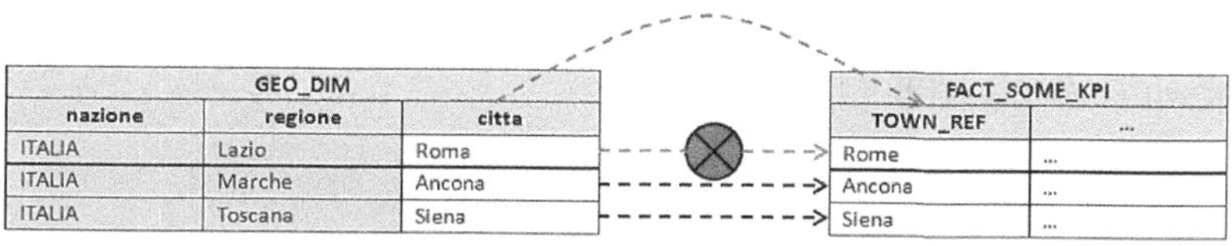

Figure 12.12. *Translated data example 1*

12.3.3. *ETL job building*

Building a job with RDM-EDE is no different than with the off-the-shelf TOS application. The full power of the standard IDE is at the disposal of the user, and this also means that in RDM-EDE it is still possible to develop non-BIVEE jobs.

RDCS extensions make the difference when it comes to BIVEE integration. In section 12.3.1, we described how meta-data in RDM-EDE are automatically derived from KPIOnto. Here, we explain how these meta-data can be exploited in job design, and we introduce a custom data output component that enables data loading on the BIVEE platform through RDM-WS.

Designing an ETL job is a task that can be decomposed in steps (see Figure 12.13):

1) Define the extraction logic: how to get all relevant data from the source system(s).

2) Define the transformation logic: how to put extracted data in the desired form. This step is optional, as the extraction logic already in place may be enough.

3) For each stream of KPI data produced in output:

i) Select the appropriate KPI schema: each data stream must be in the scope of one specific KPI.

ii) Connect the data stream to a `tRDHOutput` RDCS custom component.

iii) Adapt the schema of the source data to the KPI schema.

Points 1 and 2 are not BIVEE-specific and are handled as in any other TOS job, so they are not detailed here. The last block is actually a cycle, but for the sake of clarity let us assume that we are implementing an ETL job to deal with a single KPI.

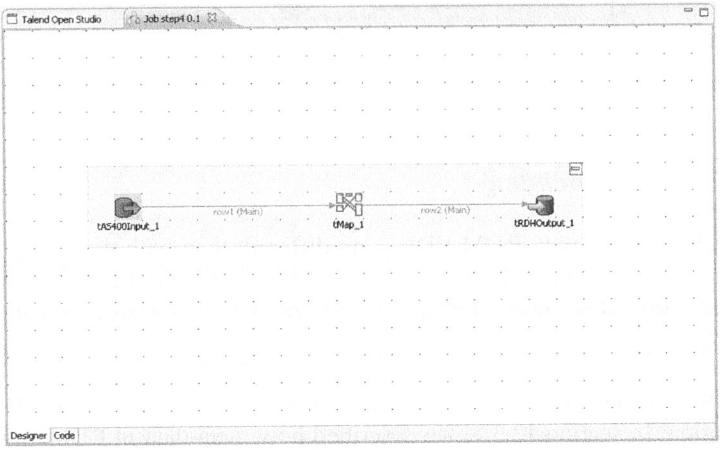

Figure 12.13. *Example of ETL job in designer*

Point 3(i) is as simple as its description: when creating the data flow toward the output, choose from the meta-data repository the KPI schema that represents the target KPI.

Point 3(ii) is straightforward: once connected with a KPI data stream, the `tRDHOutput` data output component is responsible for applying data translation rules, formatting KPI data and uploading them to RDM-WS. Configuration is minimal, as the only required parameter is the network address of the RDM-WS endpoint. The `tRDHOutput` component can automatically recognize the target KPI from the input data stream.

Point 3(iii) is where *semantic annotation* takes place: by mapping columns of the source schema to columns of the KPI schema, we are actually annotating the source data with concepts from KPIOnto. This is done using the standard `tMap` component from TOS, as shown in Figure 12.14.

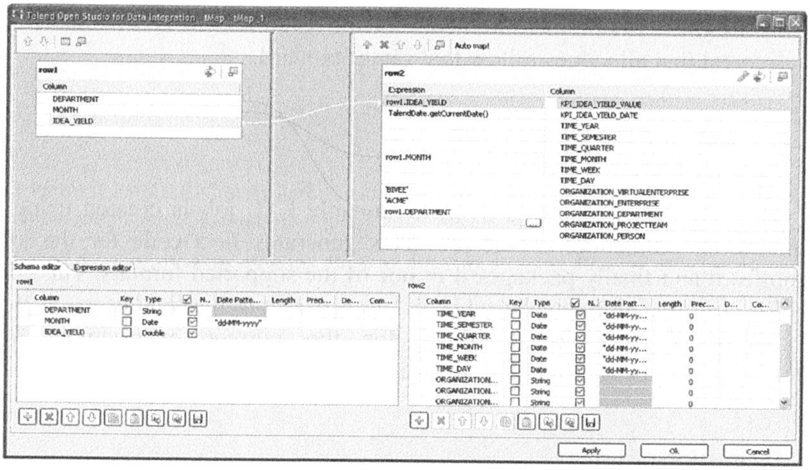

Figure 12.14. *Example of tMap configuration in ETL job*

A fundamental RDCS concept that must be taken into account when designing column mappings for an ETL job is that, as a general rule, only one single KPI column for each dimension should be mapped, leaving all the others unmapped – which means that at runtime these latter columns will always have `null` values. The reason for this is that the mapped column determines the granularity of the KPI measurements. For example, the time dimension has several levels – year, semester, quarter, month, week and day: setting a value only in the month column tells the system that the measurement is related to the given month. All the *parents* of the given month are implicit, and they are automatically inferred by the system.

The same concept applies to every other dimension that has multiple hierarchical levels, but with a caveat: if the ontology defines multiple parents for the same item, and two or more of these belong to the same level, the parent inference will fail, leading to a runtime error while loading data. To prevent these conditions and avoid ambiguities, it is allowed to map more than one column of the same dimension. For example, `OrganizationDimension` has a `VirtualEnterprise` level which should always be made explicit in KPI data, as enterprises may often be members of more than one virtual enterprise.

Finally, all KPI measurements are required to be numerical fixed point values, with a maximum size and precision of 19,6 (nineteen digits, six of which the decimal part). As the RDCS allows different data contributors (enterprises) to provide measurements for the same KPI, it is a fundamental requirement that all values follow that same standard.

12.3.4. *ETL job deployment*

Once the design phase is over, a runnable ETL job is created using a standard TOS function: export job. This generates Java code for the job, compiles it and finally packages it in one of the supported formats. The only BIVEE customization introduced here – transparent to the final user – is that all vocabulary mappings defined for the current environment are automatically packaged with the software.

12.4. Bibliography

[KIM 02] KIMBAL R., ROSS M., *The Data Warehouse Toolkit: the Complete Guide to Dimensional Modeling*, 2nd ed., John Wiley & Sons, 2002.

[ISA 14] ISAJA M., FANTINI A., The BIVEE environment prototype v3.0, Deliverable D6.33, BIVEE European project, 2014.

[TAG 12] TAGLINO F., SMITH F., PROIETTI F. *et al.*, Production and innovation knowledge repository, Deliverable D5.1, BIVEE European project, 2012.

[TAG 13] TAGLINO F., SMITH F., DIAMANTINI C. *et al.*, Advanced interoperability services, Deliverable D5.3, BIVEE European project, 2013.

PART 4

Concrete Experience of Innovation in a Knowledge Centric Economy

13

Innovation and Production Improvement in Virtual Enterprises: the User Perspective

13.1. Why validation

Pilot application validation is a crucial step for moving toward the validation of the software produced by business innovation in virtual enterprise environment project (BIVEE). Pilot application validation ensures that the BIVEE Environment actually meets the end-user's needs, and that the platform has been developed according to the user requirements. After the delivery of the first deployable version of the platform, the defined deployment strategy and pilot application validation cases have been executed as much as possible.

The BIVEE Environment is composed of several different components, each addressing different kinds of requirements of the end-user organizations. Software components of the BIVEE Environment expose a seamless integration and interoperability among each other such that the end-users are dealing with a single integrated system. Hence, during the pilot application validation process, end-users use a single platform through unified interfaces.

13.2. End-users

This section provides a description of the AS-IS status of the two end-user organizations in BIVEE, namely AIDIMA and Loccioni. The AS-IS

Chapter written by Anil PACACI, Ali Anil SINACI and Asuman DOGAC.

status presents the innovation and production operations of both end-users before the introduction of the BIVEE Environment. The activities that were carried out inside the organizations concerning innovation management, production management and information management are described in detail through information flow analysis diagrams and user specification tables.

The AS-IS status of the end-user organizations is analyzed through two main topics as presented in Figure 13.1:

– Information flow analysis intends to give detailed information about the main processes of the organization. In this part, process flowcharts and their descriptions are presented conceptually.

– User specification provides information about the main actors of the activities, their responsibilities and roles within the processes. This topic also gives conceptual information about the users and their roles. Conceptual users and their associated roles are presented in a user specification table. This table is the basis for the analysis of data exchange between users.

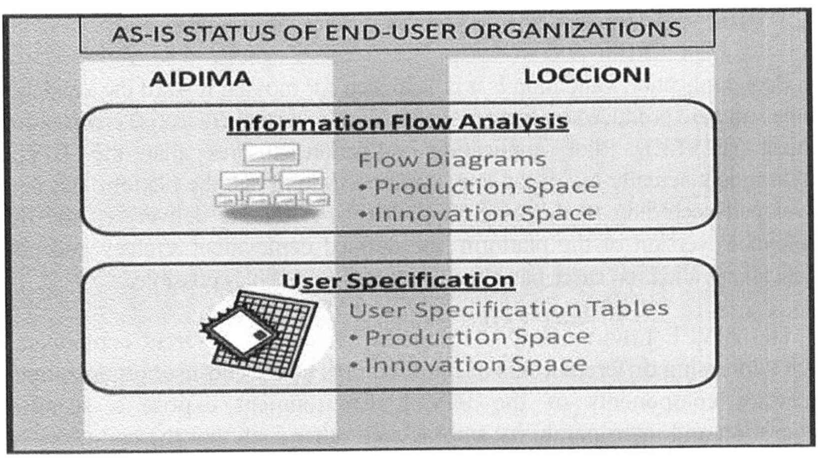

Figure 13.1. *Document-centric AS-IS analysis*

13.2.1. *AIDIMA*

AIDIMA is a research association in the area of furniture business. It is not a manufacturer, but remains at the core of the furniture companies when an innovation is in action. Innovation processes are not well defined and follow an informal way through phone calls and *ad hoc* meetings. AIDIMA organizes the collaboration among different furniture-related organizations and no special software is in use covering the whole process. Most of the furniture manufacturing companies do not make use of informations and communications technology (ICT) tools either, in any decision-making process.

For AIDIMA, BIVEE seeks to build the necessary software systems to improve the ICT adoption among furniture companies, mostly on traditional production/logistics chain. Integrating ICT tools addressing the exchange of important information between the partners and advanced software systems for the innovation processes of AIDIMA is another important goal in the case of this end-user company.

13.2.2. *Loccioni*

General Impianti (Loccioni) is a system integrator with a core focus on the development of innovative "customized" solutions to solve specific problems for the improvement of quality, efficiency and sustainability of products and processes. Loccioni is the end-user organization through which BIVEE capabilities, mostly in innovation management, will be validated.

The end-user works with several different external partners in accordance with the Open Innovation scheme. Loccioni shares the innovation processes with external entities, such as research centers, universities, suppliers and retailers. Since Loccioni is a specialized company in innovation, it makes use of several ICT tools during different phases of innovation and later production. Several departments within the company work together. Coordination among the departments and executives of the company is of crucial importance in creating the correct results which will lead to correct decisions.

Loccioni has well-defined processes for innovation management; however, there are no specialized and integrated software tools for the document exchange in a semantic and formalized way. There they remain at a conceptual level on the innovation management. The company knows what to do when an employee comes up with an innovative idea, or when managers want to implement an idea. However, these processes are not ICT-enabled as a whole. Several specific software tools are in use throughout the company, such as Oracle JDE (J.D. Edwards) as an enterprise resource planning (ERP) or an in-house developed people relationship management (PRM).

13.3. Pilot validation cases

The BIVEE Environment has the objective of considering the management of business innovation and the improvement of production processes. Hence, the application cases are organized into two sections: innovation space for business innovation and production space for improvement. The innovation space section describes the processes involved in the innovation management while the production space section presents scenarios related to manufacturing and production processes.

The goal of this section is to provide the definition of detailed scenarios and pilot application validation cases in order to achieve the user validation of the BIVEE Environment. These application cases are defined by General Impianti – Loccioni (LOC) and AIDIMA, representing the end-user organizations of the BIVEE Environment.

In the previous section, we described the end-users and the current AS-IS situation of both companies in order to understand the tasks they perform in their organizations. These AS-IS scenarios and descriptions present the operations performed by both companies in the innovation and production spaces. Although both companies have competences in innovation and production areas, Loccioni has more experience in innovation activities while AIDIMA is more focused on production processes as described earlier. To be in line with the focus of each end-user, the scenarios proposed by Loccioni will be focused on the innovation space, whereas the AIDIMA scenarios will be focused on the production space.

For the development of the innovation space scenarios, the current state of innovation management in Loccioni has been analyzed, identifying the most relevant processes and the main actors involved. This is done almost internally by each company. Some interviews with innovation management staff from each company have been undertaken. Through these meetings, the processes, actors and documents involved have been identified. The same methodology has been carried out during the development of the production space scenarios with some differences. In order to define suitable production cases, some meetings have been arranged with external companies and customers of AIDIMA. The situation and difficulties of each company have been evaluated in order to find the most interesting application cases based on the expected functionality of the BIVEE Environment.

All the pilot application scenarios are defined on virtual enterprises; therefore, they imply "cross-enterprise" processes with several enterprises involved. The selected scenarios are not composed of only the optimal cases; they have been selected in a realistic perspective. We expect that they indicate possible problems that may arise during the value production processes, such as delays and faulty production. This ensures that monitoring capabilities, support to exception handling and contingency planning functions of BIVEE are extensively tested during the validation process.

13.3.1. *AS-IS and TO-BE application cases*

To come up with the pilot application cases for the innovation and production spaces, two kinds of scenarios are described: AS-IS and TO-BE application cases. The AS-IS scenario represents the current work flow of the company while the TO-BE scenario describes the "desired" work flow focusing on the problems found in the current processes and trying to avoid them. In addition, in the TO-BE scenarios, any kind of optimization on the processes is considered carefully. Both scenarios describe the processes performed inside each scenario, the actors involved in these processes, the documents generated and consumed, and the technology employed during the execution of these processes.

The methodology used for the definition of the TO-BE scenarios is based on the analysis of the current company processes (AS-IS), the evaluation of these processes and the detection of the problems, needs or improvements that can be applied to these processes. Although both AIDIMA and Loccioni contributed to the development of the scenarios for both innovation and production spaces, pilot application validation is planned to be performed in one space for each end-user organization. In line with BIVEE objectives, Loccioni is planned to validate the innovation space and AIDIMA is planned to validate the production space.

The scenarios defined by end-users for both innovation and production spaces are in line with the business innovation models and methods that have been defined within the scope of BIVEE business innovation models and methods. Pilot application validation cases are strongly linked to the key performance indicators (KPIs) which have been identified within the same models and methods. Real usage patterns of the KPIs are reflected within the validation cases. The first measuring campaign is dealing with the associated measurements of these KPIs within the defined AS-IS application cases.

13.3.2. *Connection to user requirements*

The process of requirements gathering and analysis can be complex and arduous. Users do not often understand their systems, let alone have objective views of how they work currently or should work in the future. Fortunately, we have been assisted by the BIVEE representatives in AIDIMA and Loccioni which have tried to shield the project from these difficulties, although we should recognize the effort it has taken.

Pilot Validation Cases task has a close collaboration with the Analysis of User Requirements. The reason is that the nature of requirements analysis is closely related to the definition of AS-IS and TO-BE scenarios. In the following, the produced content is presented through a time line:

– The descriptions of AIDIMA and Loccioni are first analyzed and presented. These descriptions present the current activities through a high-level window and they are based on interviews with the end-user representatives and the questionnaire analysis. The questionnaire has 29 hierarchically designed questions, mostly requesting information about the innovation activities, as presented in Table 13.1. Having analyzed the answers to the questionnaire, we came up with a detailed analysis of AIDIMA and Loccioni. The main objective is to understand the current business domain, business models, production activities and the way these organizations look at the concept of innovation.

– A comprehensive AS-IS analysis has been performed. This technical work is performed through information flow analysis and user specification tables. This work follows the document-centric approach [SIN 12].

– Analysis of the AS-IS status of the end-user organization has led to the problems which exist in the current processes. Analysis of the identified problems has been performed within the Analysis of User Requirements.

– Based on the analysis of the AS-IS status and identified problems, the TO-BE descriptions of the end-user organizations have been presented. This is a high-level description for the expected functionalities of the BIVEE platform from the end-user's perspective. Further analysis has been performed to provide the formal user requirements and use-case definitions.

– To come up with the accurate KPI definitions, before the TO-BE scenarios, AS-IS scenarios are presented under the same sections belonging to corresponding end-user. Scenarios are called as "cases" within the deliverable to reflect that they are based on the deployment and validation of the expected BIVEE platform.

Table 13.1 presents the questionnaire, which has been applied to end-user organizations, in a hierarchical way.

No.	Question
1	How do you define a product in light of the discussions made in the first plenary meeting?
1.1	Based on your definition, does your company have any products? What are they?
2	How do you define innovation?
2.1	Could you give some specific (imaginary, maybe impossible to achieve) examples?
2.2	How do you define innovation in the context of your company? Any examples?
3	What is the innovation management process of your company?
3.1	Do you have a separate R&D department to manage innovation within your company? If yes, provide the details.
3.2	Do your managers push ideas to the employees? If yes, provide the details.
3.3	Do you organize brainstorming sessions to create innovative ideas? If yes, provide the details.
3.4	Do you apply any performance evaluation for the production of innovation within your company?
3.5	Do you expect innovative ideas from your employees? If yes, provide the details.
3.5.1	What happens if an employee of your company comes up with an innovative idea?
3.5.1.1	How does he/she share the idea with the managers?
3.5.1.2	How does he/she share the idea with the rest of the company?
3.6	Do you share innovative ideas with your business

	partners (other companies such as providers, retailers and customers)?
3.6.1	If you share, how do you do it? Do you use any specific platform? Or phone calls? Or reports, etc.?
3.7	What actions are taken based on this idea within the company?
3.7.1	Which departments are involved in the discussion of the idea?
3.7.2	What is the methodology in this discussion? Do you have any formal processes for this purpose?
3.7.3	What happens based on the results of the discussion?
3.7.4	Do you apply any planning activities, if the idea is accepted as valuable? If yes, what are the details?
3.7.5	Do you apply any forecast or monitoring activities, if the idea is accepted as valuable?
3.7.6	Is there any awards for the employee who came up with the idea?
3.7.7	What are the equivalent actions if the idea is not accepted as valuable?
4	Do you have specific programs/activities to increase the innovation capabilities of your employees? If yes, what are these activities?
4.1	Who can attend these activities? Any selection criteria?
4.2	What is the frequency of these activities?
4.3	Are there any specific tools to organize and manage these types of activities?
4.4	Do you have any specific action/process to lead from these activities to innovative ideas? If yes, what are the details?

Table 13.1. *Questionnaire for the end-users*

13.4. First monitoring campaign (FMC)

The goal of a monitoring campaign in an enterprise is to determine the current condition of the enterprise through values of a set of variables. Adopting KPI for monitoring is one of the commonly used approaches in today's organizations. KPIs, sometimes called "performance indicators", are the critical measures that focus on organizational performance for the current and future success of the organization [CUS 10].

During the FMC, the main focus was on the AS-IS status of the end-users by using a selected set of KPIs and drawing logical assessments out of them. As expected, in the second monitoring campaign (SMC), the focus is on the status of end-users with a BIVEE platform in action in their environments. Hence, the KPI measurements let the consortium understand and show the impact of the BIVEE platform. In other words, KPIs are the means to show the impact within the 40-month project lifetime. Monitoring KPIs with and without a platform for a period and comparing the well-understood KPI values are the core of the validation activities of the BIVEE project.

13.4.1. *Methodology*

For the technical scope, AIDIMA and Loccioni have presented a number of projects in D7.11, which are appropriate for the goals of the BIVEE project and which carry the notion of virtual enterprises. Each end-user defined two pilot application cases. For the FMC activities, we have selected the ones that have wider coverage in terms of collaboration, activity and innovation and improvement intensity.

AIDIMA defined two different pilot application cases. Both application cases share the most important key points, but the first one (chair by Muebles Romero) covers a wider application scope than the second one (kitchen by Gellida). We can consider that the first scenario includes more collaborative processes regarding the production of some special or customized goods while the second scenario is more focused on internal processes and partners.

The first pilot application case, based on the production of the chair 145, presents some very interesting features. The manufacturing of this model of chair requires the collaboration of different external partners of Muebles Romero in order to obtain not only the materials but also the production of some product parts and the application of some finishes which can be only covered through external partners and should be considered as external operations.

In the meantime in Loccioni, for the FMC, it has been decided to monitor the following types of activity/project:

– a project devoted to the development of a new system, named Felix;

– new ideas proposed.

The Felix project has been considered as innovative and is developed by the Research for Innovation (RforI) team because it is related to an application of core competences of the RforI team (i.e. machine vision and robotics) in one sector (the railway infrastructure) not covered by any of the Loccioni business units.

Name	Description
Reliability	The objective of constantly achieving operative goals
Velocity	The objective of fast reaction and short cycle times
Adaptability	The objective of being able to adapt to varying external or internal demands
Cost orientation	The objective of operating on minimal costs
Asset orientation	The objective of generating tangible and intangible profit
Innovative potential	The objective of successfully carrying new ideas to the market
Customer orientation	The objective of satisfying customers
Network orientation	The objective of a maximal efficient network

Table 13.2. *Summary of business objectives from D2.2*

The scope of monitoring campaigns does not only cover KPI measurements, but also presents an assessment of the monitored values. The assessment has been performed according to a number of profiles. A profile, in this respect, is the interpretation of a group of KPIs with a specific goal and their measurement values. Business objectives of our end-users are used as profiles to give a more structured notion to the analyses. These business objectives are formally defined in "The BIVEE Framework and the Collaborative Innovation Capability Maturity Model". In Table 13.2, an overview of the business objectives of AIDIMA and Loccioni is presented. We evaluate the measurements and KPIs in terms of these objectives in section 13.4.2 – Anaylsis and Conclusion.

Systematically, each KPI relates to a business objective. In other words, a business objective is defined by the collection of a number of KPIs. The values for the KPIs determine how successful the company was about the specific objective.

13.4.2. *Analysis and conclusion*

In this section, results of the FMC in terms of KPI measurements are analysed seperately for each end-user organization. High level assessment is presented about the status of the end-users by the end of FMC.

According to the results of measurement and monitoring activities conducted during FMC in AIDIMA, the following inferences are made:

– focus on internal operations, 67% of the indicators are internally focused;

– not human centric, only 4 indicators out of 52 are human centric;

– focus on monitoring production rather than innovation or improvement.

Figure 13.2 shows the business objective profile of AIDIMA according to measured KPIs.

Similar inferences made about Loccioni after FMC are as follows:

– In total, Loccioni has also focused internally on monitoring mechanism. This situation establishes a goal for the BIVEE project, increasing the focus on external activities, especially in the creativity wave.

– 10% of KPIs have an obvious human focus. This is due to the fact that the KPIs used in the FMC are more related to business objective and research project, not the people.

– The business innovation space KPIs are, of course, all innovation related, all concentrating on the management of ideas through the pipeline from creativity to implementation.

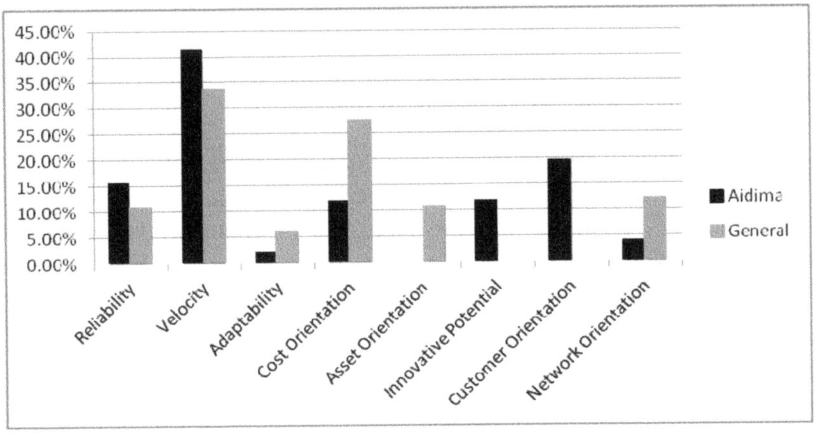

Figure 13.2. *Concentration of furniture companies and general production related KPIs*

Figure 13.3 presents the Loccioni profile according to business objectives to complete higher level analysis of FMC.

The starting point in the monitoring campaign was to analyze the current condition of the project. This analysis focused on the business of end-users and already available material produced by their pilot application cases reflecting the AS-IS status. We investigated the methods of collecting performance measurements, and analyzed the business of end-users and the reasons behind our methodology, the scope of the monitoring and the process followed during the FMC.

KPIs are numeric indicators and on their own are important information sources for the enterprises. FMC is structured around KPIs. However, the real value comes with the meaning behind the numeric values. It is needed to assess an increase or decrease in a value of KPI to see the real implication of the change. In this respect, we focused on constructing a number of profiles out of the KPIs and came up with some assessment.

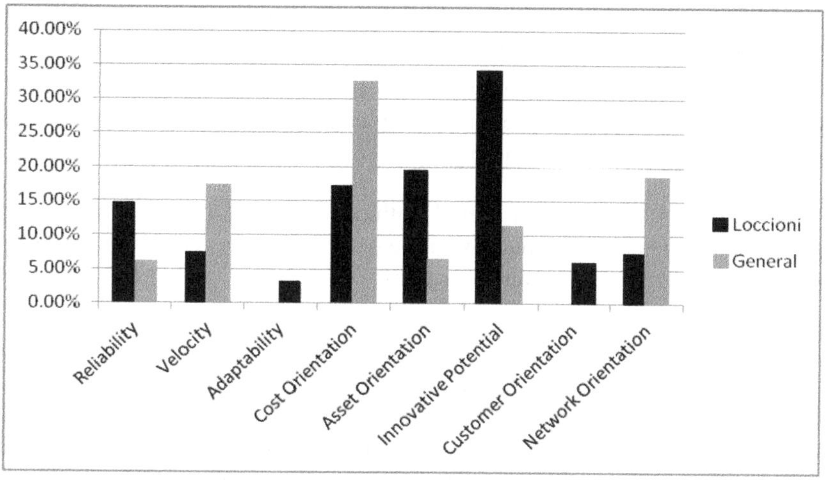

Figure 13.3. *Concentration of Loccioni and general innovation related KPIs*

We also gathered tacit knowledge about the business of the associates of the AIDIMA and Loccioni. It is also likely to observe improvements on these concepts with the deployment of the BIVEE platform. Below, there is a list of lessons and knowledge learnt during the campaign:

– Monitoring activities require a great deal of *communication*. BIVEE may increase the efficiency on the monitoring activities.

– If it is not recorded, it is difficult to collect data for the previous periods. BIVEE may be helpful with the previous records.

– Monitoring of separate departments of the same enterprise is a cumbersome process because of the overlapping indicators and dependencies.

– It is even harder to monitor an enterprise within a virtual enterprise. An indicator may depend on many cross-enterprise factors.

– High frequency of monitoring activities is not helpful for virtual enterprises. BIVEE may be helpful by providing mechanisms to handle the frequency and to retrieve more information with the same amount of time.

– Weekly meetings for monitoring activities are too ambitious. This knowledge will be helpful during the SMC.

– Businesses change every day. In particular, the bigger the enterprise is, the faster it changes. Thus, *change management* is important.

– Having available software solutions greatly eases the required effort. Without a dedicated software managing data, e-mails and other previous software records are not very helpful.

These points are also considered within other validation activities since this information is helpful not only in the next monitoring campaign, but also in the requirements analysis and scenarios.

13.5. Second Monitoring Campaign (SMC)

In the SMC, an analysis similar to the FMC is performed. The difference lies in the deployment of the BIVEE Pilot Application at the end-user settings. The value for the selected KPIs, similar profiles and similar high-level assessments is discussed during the SMC.

The performance information gathered before and after the introduction of BIVEE Environment at the end-user settings realizes a comparative analysis. This plays an important role in the validation activities of the BIVEE Environment together with the Pilot Application Validation Cases.

13.5.1. *Methodology*

According to the definition of Task 7.3, SMC is defined as "running the validation studies of the BIVEE pilot applications that are enabled on top of the Integrated BIVEE Platform and realizing the second systematic monitoring campaign (KPI-Ph2) on two BIVEE Pilot Applications with the improved set of KPIs used in the first campaign, for the BIVEE supported value production and innovation management processes".

Within this monitoring campaign, due to the shift in the BIVEE methodology through the cooperation of end-users and BIVEE Environment, slightly improved versions of the TO-BE scenarios have been used to monitor end-users' production and innovation capacities in order to perform an optimal evaluation.

During the SMC, it has been seen that there is a continuous interaction between the platform and its users. Apart from the KPI values, the BIVEE Environment does have important effects on the culture of innovation of the end-user enterprises. On the production site, BIVEE enabled small and medium-sized furniture enterprises (SMEs) to monitor their status and provide infrastructure that makes collaborative production easier. On the innovation side, BIVEE provides improvements over the existing innovation methodologies and gives necessary toolset to make innovation management easier and efficient. As a result, end-users provided continuous feedback for the continuous improvement of the BIVEE Environment since they experienced the platform from the beginning of the SMC. This enabled technical partners to work on detected issues and possible improvements and see the user reaction in a much faster manner. To summarize, the execution of monitoring campaigns has changed their natures. It was not only measuring but also cooperating with the technical partners in order to improve both environment and end-user processes. In order to make the most benefit out of cooperation, SMC and respective tasks have been extended until the end of the project.

The monitoring process of the SMC is similar to that used in the first campaign, except for the fact that the BIVEE Environment is in use. One difference is that the analysis of the SMC has been held in two phases in order to monitor the impact of cooperation. Another important difference is to have a formal quality evaluation of the BIVEE platform according to ISO SQuaRE [ISO 11] standards, which presents a formal software validation methodology.

Validation of the software product could be considered as answering the following question: "are we building the right product?" [BOE 89]. The objective is to understand whether the software meets the used needs and requirements were correct in the first place. BIVEE validation and verification activities use the Software Product Quality Evaluation Reference Model as a reference, which describes the process, activities and tasks performed during the quality evaluation of a software product (Figure 13.4) and provides a quality model to evaluate the quality of software product and effect of the software product on end-users.

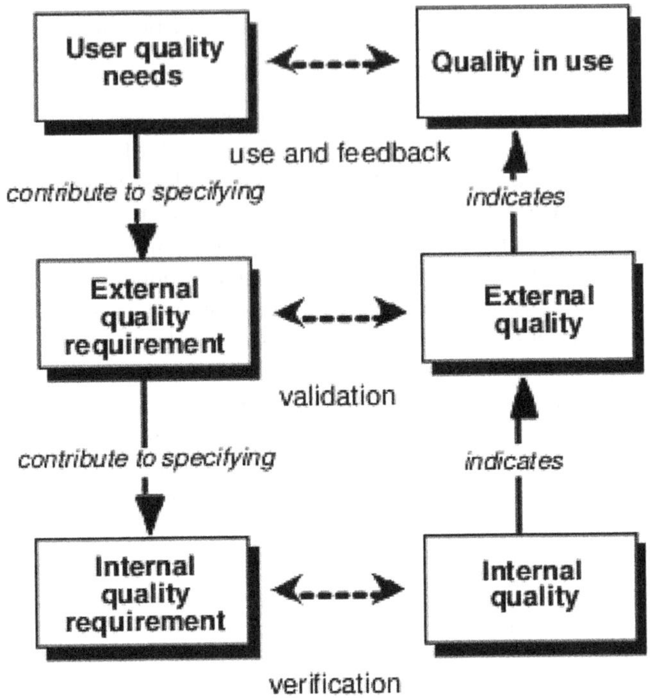

Figure 13.4. *Stages of software lifecycle*

Evaluation of the internal characteristics is usually carried out during the development process to assess the quality of the intermediate products. Therefore, internal quality is out of the scope of this chapter.

External quality is the totality of the characteristics of the software product from an external view when the software is executed typically in a simulated environment. According to Figure 13.4, measuring the external characteristics is considered as the validation activities. In a more general context, validation has to ensure that software meets the user needs and functions as expected.

Two different sets of surveys have been prepared for the validation activities based on the evaluation design: developer surveys and tool-specific surveys. The distinction is based on the target set of users and eventually the quality characteristics.

The developer survey focuses on the reliability, maintainability and portability quality characteristics. Together with the requirements survey, they function as the software validation of the BIVEE Environment Validation. Another set of surveys has been prepared to evaluate functionality, usability and efficiency characteristics. This time, instead of one single generic survey, different surveys have been implemented for each BIVEE component. The result of these surveys reveals the external quality of the BIVEE components in terms of the end-user point of view. It helps us to understand whether the current platform is sufficient to meet user needs and whether each function provided by BIVEE tools works as expected by the end-user.

To evaluate the quality in use which comprises four quality characteristics, a separate survey is prepared. This survey includes 15 questions. Five of the questions are intended to evaluate effectiveness and productivity characteristics of quality in use model. The rest of the questions are obtained from the standardized questionnaire SUS [BRO 96]. The system usability scale was first released by John Brooke in 1986, originally created as the "quick and dirty" scale, which is an evaluation technique based on evaluation user experience in daily use, for a usability test. Its technology-independent nature has made it a widely used usability standard in measuring end-user satisfaction on usability. SUS is also one of the SQuaRE recommendations as satisfaction metrics.

13.5.2. *Analysis and conclusion*

This section presents the results obtained from the two phases of the second systematic monitoring campaign. Comparison of these results with the previous one is going to be the basis for the BIVEE impact analysis.

It is important to assess cultural effects of the BIVEE apart from the numerical changes in KPIs. After a 4-month period of SMC, it has been observed that the BIVEE Environment itself adds a big value, especially for the production space. Traditional furniture companies do not have a system to monitor their processes. On the other hand, companies in the innovation space usually do not have a defined methodology for innovation management. Assessments of end-users indicate that BIVEE has been successful in addressing these two problems by providing a methodology and a service platform both for process monitoring and innovation management.

The 4-month experience of the first phase of the SMC has produced valuable feedback for the improvement of the BIVEE platform. Technical partners have already been provided with this feedback during the SMC in an informal way by e-mails, issue tracking systems, dedicated conference calls, etc. More effort has been made to evaluate the end-user experience in a standardized and systematic way. For this purpose, a number of surveys and analyses based on ISO/IEC SQuaRE have been prepared and conducted. Apart from some minor issues, it can be concluded as a result of validation activities that BIVEE Environment has reached a satisfactory level of external quality and quality in use. To be able to further improve adoption of the BIVEE Environment both in production and innovation spaces, the following proposals have been underlined by the end-user for the final version of BIVEE platform due to the second phase of the project:

– the RDH interface provided for the manual submission of KPI values could be improved in order to make it more user-friendly. Some functionality such as searching or filtering KPIs could greatly improve the user experience while introducing some KPI values manually into the platform;

– although the overall impression of the users about the MCR is positive, some aspects can be improved according to the retrieved feedback. They are not large alterations, and in exchange users could be able to get more out of this BIVEE component;

– a higher level of integration of the different BIVEE tools should be achieved in order to provide a unique system to the user and ensure a complete communication between the different components;

– the introduction of a fully integrated platform among the different developed tools in order to facilitate the adoption of the whole platform;

– the fully automatic management of the different KPIs through VIF and BIVEE EDE in order to evaluate the tangible and intangible benefits during each wave;

– the adoption of a more user-friendly interface for the KPI monitoring inside the VIF.

Until the end of the project, the BIVEE Environment has continued to improve through end-user feedback and has taken its final shape at the end. During the second phase, similar monitoring activities have been performed. Further adoption and improvement of the platform has enabled us to collect more quality data during this period, which results in more significant changes in observed data. Apart from monitoring activities, analysis conducted to evaluate BIVEE software quality has also been performed during the second phase to assess the final version of the BIVEE Environment. This helped us to observe improvement of the BIVEE Environment in a systematic and standardized way.

Evaluation of the validation surveys and end-users' analysis have revealed that end-users' experiences and feedback have been taken seriously by the project to give the platform its final shape. In the first phase, end-users have reported issues on the usability of the tools, and the SUS scores obtained from validation surveys were lower than those expected. During this last period of the SMC, technical partners worked to improve this usability issue, especially on RDH. The second round of SUS study shows a dramatic increase in the usability of the RDH and other BIVEE tools. Apart from that, end-users suggested new functionalities and improvements over the existing ones. Specifically, Virtual Innovation Factory has realized many important features suggested by end-users, such as RDH integration to collect KPI and a user-friendly interface for KPI monitoring.

13.6. Impact analysis of BIVEE

One of the most important aspects of the SMC is the continuous interaction between the platform and its users. In fact, having the BIVEE Environment in action during the SMC, end-users have been able to provide feedbacks about their experience both related to cultural and technical aspects. This valuable feedback has been addressed during the platform fixing and refinement activities to give to the BIVEE Environment its final shape. This enables technical partners to work on detected issues and possible improvements and see the user reaction in a much faster manner.

In this way, the methodology of the SMC has been changed because, first, the consortium understood that focusing solely on the performance measurements and forgetting about the social and cultural aspects could underestimate the BIVEE impact; then, the SMC is considered closer to co-working approaches than the waterfall model planned at the beginning of the project. Therefore, there is a bilateral improvement related to the cultural improvement that BIVEE framework provided to end-users, on the one hand, and the big enhancements introduced in BIVEE Environment due to the feedbacks collected from end-users and stakeholders, on the other hand.

On the production side, the key points obtained from AIDIMA's experience adopting the BIVEE Environment are the following:

– The MCR has provided a new interface that improves the user experience. This new interface, especially the MCR Modeler component, has been favoured by users. In addition, users have welcomed the translation of the interface into Spanish.

– The introduction of manual KPIs has been boosted by the creation of ETL jobs to read the information from an Excel file and populate it to the platform by using the BIVEE ETL Development Environment.

– The BIVEE Environment has achieved a higher level of integration providing a unique access sign-on point to access the different BIVEE tools.

– The list of KPIs has been revised again in order to organize them and maintain a track during the population of its values.

– The RDH has finally included important improvements, which makes the introduction of new KPI values easier to the users.

– Most of the companies from the furniture/wood sector are reluctant to share some information about their KPIs with the rest of the VE members. This barrier has been addressed to provide useful KPI values for the SMC activities.

On the other hand, the Loccioni experience points out that:

– The selected KPIs have been revised and separated into two main categories: KPIs able to describe the innovation capacities of Loccioni at business innovation space level and KPIs related to a single project linked to the virtual enterprise. The VIF allows checking most of both kinds of KPIs: those related to the innovation business objective that can be visualized in KPIs drop down menu item and those related to innovation project that can be visualized in the corresponding tab in the specific innovation project page.

– From the FMC starting to the end of SMC preparation phase, the management of the proposed ideas has been implemented and it is possible to monitor the percentage of ideas accepted and the quality in terms of completeness of information provided.

– During the second phase of SMC, the automatic collection of most of the KPIs has also been implemented through the VIF for the Network, Reliability and Asset orientation business objectives. In this way, the analysis will be completed and effective for the strategic management of the innovation.

– A higher level of integration among the different developed tools in order to facilitate the adoption of the whole BIVEE Environment is completed during this second phase of SMC.

– The automatic management of some KPIs is available from VIF and RDH, in order to evaluate the tangible and intangible benefits during each wave together with the insertion of a more user-friendly interface into the VIF for the KPI monitoring.

– These tools and methodology are very useful in order to avoid sending e-mails with big attachments and to have a common space to share useful documents. The sharing of comments within an innovation team is also very effective in terms of collaboration speed. Improvement in the understanding of all the reports and KPIs leads to very successful management of innovation projects.

Chapters 4 and 5 present two typical case studies of production and innovation cases with the introduction of the BIVEE Environment. The two pilot cases are examples of the improvement of the two end-users environments due to the use of BIVEE Environment.

During the description of the pilot cases, all the improvements introduced by BIVEE Environment are highlighted. As already discussed, the evaluation of the improvement has not been done on the basis of KPI evaluation because it is not possible to compare the benefits using different projects. In fact, mainly for the innovation case, the pilot case chosen is not the same used in the FMC. The comparison and benefits are described in terms of improvements in the management of production and innovation and also in terms of cultural/management changes.

Due to all the BIVEE tools, the monitoring of KPIs and the availability of schema and documents, the decision is simpler and the interaction among the members of the VE is very useful and fast. Through the BIVEE Environment, all the steps quickly proceed and with an automatic tracking of all the interactions occurred in both innovation and production spaces.

From the end-user's point of view, the BIVEE Environment with its tools, and in general the BIVEE project, has been a useful approach and a technological boost in order to bring innovation into a more traditional sector, such as the furniture sector, and to accelerate the innovation process management in an innovative company.

13.7. Bibliography

[BOE 89] BOEHM B.W., *Software Risk Management*, IEEE Computer Society Press, 1989.

[BRO 96] BROOKE J. "SUS: A quick and dirty usability scale", in JORDAN P.W., THOMAS B., WEERDMEESTER B.A., *et al.* (eds), *Usability evaluation in industry*, Taylor & Francis, London, 1996.

[CUS 10] CUSUMANO M. A. Staying power: Six enduring principles for managing strategy and innovation in an uncertain world (lessons from Microsoft, Apple, Intel, Google, Toyota and more). Oxford University Press, 2010.

[ISO 11] ISO/IEC 25040:2011, Systems and software engineering – systems and software quality requirements and evaluation (SQuaRE) – evaluation process (n.d.), available at http://www.iso.org/iso/catalogue_detail.htm?csnumber=35765, 27 May 2015.

[SIN 12] SINACI A.A., PIERSANTELLI M., CRISTALLI C. *et al.*, "A document centric approach for user requirements in BIVEE", *New Generation Enterprise and Business Innovation Systems*, 2012.

14

A Methodology for the Setup of a Virtual Innovation Factory Platform

14.1. Introduction

The business innovation in virtual enterprise environments (BIVEE) aims at deploying advanced methods for boosting creativity and innovation with an open innovation approach, for supporting their lean implementation and for monitoring the concrete outcomes, managing innovation ventures in a collaborative networked industrial setting.

The open innovation model is theorized by Henry Chesbrough, who is considered to be the father of this approach; in open innovation, the new imperative for creating and profiting from technology [CHE 03]: "is the use of purposive inflows and outflows of knowledge to accelerate internal innovation, and expand the markets for external use of innovation, respectively. This paradigm assumes that firms can and should use external ideas as well as internal ideas, and internal and external paths to market, as they look to advance their technology". Open innovation implies to gain access to distributed knowledge. New value is created from a new innovation environment, composed of people inside and outside the company. It is a way to accelerate innovation, and it can give the capabilities to improve velocity (time to market reduction), direction and new business paths.

An important aspect is to establish a methodology to quantify open innovation levels. There are a lot of existing works in the literature that try to define the concept of open innovation, but there are also several discussions

---

Chapter written by Cristina CRISTALLI, Daniela ISIDORI and Isabella TERZONI.

and methods that suggest the development of the open innovation measure based on a diverse set of activities [RAN 13, SCH 11], for example the degree of openness through the number and type of partners and the phases of the innovation process [LAZ 10]. Rangus *et al.* [RAN 13] defined the tendency for open innovation as "the firm's predisposition to perform open innovation activities, such as external participation, inward Intellectual Property (IP) licensing, external networking, outsourcing of R&D, customer involvement, employee involvement, venturing and outward IP licensing". These aspects could become key performance indicators (KPIs) for the measuring of open innovation level of an enterprise.

The members of a virtual enterprise (VE) in a business ecosystem (BE), working on open innovation activities, require a continuous interaction because they have the intent to share common resources to do business together. Therefore, the management of the VE (i.e. removal of an enterprise from a VE, contact details) is a key aspect and it is handled through the BIVEE environment [DEP 12].

For this reason, the introduction of a platform able to manage the ideas and follow their development is needed. BIVEE addresses the full lifecycle of the "inventive enterprise" that spans from creativity and idea generation to engineering of new businesses to continuous production improvement. The BIVEE platform is based on a mission control room (MCR), for monitoring of VEs value production activities, of a virtual innovation factory (VIF), for continuous innovation creation and of an advanced knowledge repository as the production and innovation knowledge repository (PIKR). At the same time, BIVEE provides means for measuring the success of any process improvement or innovation venture (Raw Data Handler – RDH)[1,2]. For references, see Part 3 of this book.

Through the VIF tool, users can propose their ideas, update their proposal and inform themselves about other idea submissions. These submissions can have technical materials attached. Feedback mechanisms such as comments are very useful. Validation shows itself in every domain in different levels. Before the idea proposals, validation mechanism is published and this is conducted through a domain expert. The position of domain expert is

1 BIVEE deliverable D2.2 – specification of business innovation reference frameworks (in the context of the VEMF), see http://wordpress.bivee.eu/resources/publications/.

2 BIVEE deliverable D6.12 – analysis of user requirements, see http://wordpress.bivee.eu/resources/publications/.

important because it includes the call for ideas for pulling innovation and should evaluate the inserted issues in order to eliminate irrelevant ideas.

The BIVEE environment has been tested involving end users with two pilot applications from different domains, addressing a spectrum of cases that involve different degrees of innovation activities. As end users of the project, Loccioni and AIDIMA are involved in the definition of the pilot application scenarios for mapping the requirements of the pilot applications to the actual architecture of BIVEE. BIVEE environment validation is responsible of running validation studies of the BIVEE pilot applications enabled on top of the integrated BIVEE environment, and for this purpose it has realized the second of two systematic monitoring campaigns.

The first monitoring campaign has the objective of monitoring the AS in the context of the AIDIMA and Loccioni for building a profile of their production and innovation capabilities structured according to the performance indicators. It has given the first set of KPI values to be used as reference values during the second monitoring campaign (SMC) and, finally, for the impact analysis. During the first monitoring campaign, it has been understood that focusing solely on the performance indicators without considering the cultural and social aspects of the BIVEE environment could be misleading and not representative of the innovation aspects involved in the project. In order to better analyze the consequence of BIVEE environment use on the innovation culture of the end users, it has been decided to have a more qualitative analysis with respect to a pure quantitative analysis based on KPIs. Apart from monitoring the end users' processes on the deployed pilot applications, one of the most important objectives of the SMC is to provide feedbacks for the platform enhancements. During the SMC, Loccioni, as end users, and its related stakeholders are called to provide feedbacks regarding any issues detected with the implementation of the VIF. In this context, the support of technical partners for specifying the requirements of the actual architecture of BIVEE is a fundamental aspect[3].

14.2. Innovation knowledge flow, storage and monitoring with the BIVEE platform

The proposed innovation flow is organized into four (partially overlapping) parts, which can be referred to as waves [DEP 12] that evolve

3 BIVEE deliverable D7.40 – comparative impact analysis of BIVEE pilot applications, see http://wordpress.BIVEE.eu/resources/publications/.

over time and their produced achievements are reported in a sequence of documents that follow a consistent logic. For each wave, several groups of documents and a set of dependencies among documents are proposed. The four waves are creativity, feasibility, prototyping and engineering. Creativity starts with an innovation idea or a problem to be solved, described by a number of preliminary documents and with the creation of an innovation team. Feasibility defines the impact and includes refined planning to justify the required investment. Prototyping involves the first implementation of the initial ideas, achieving a first full-scale working model that is tested and analyzed to verify the actual performance and characteristics. Engineering aims at producing the specification of the final version of the new product ready for the market, and the corresponding production process [DEP 12]. The full innovation flow from the end user's perspective is shown in Figure 14.1.

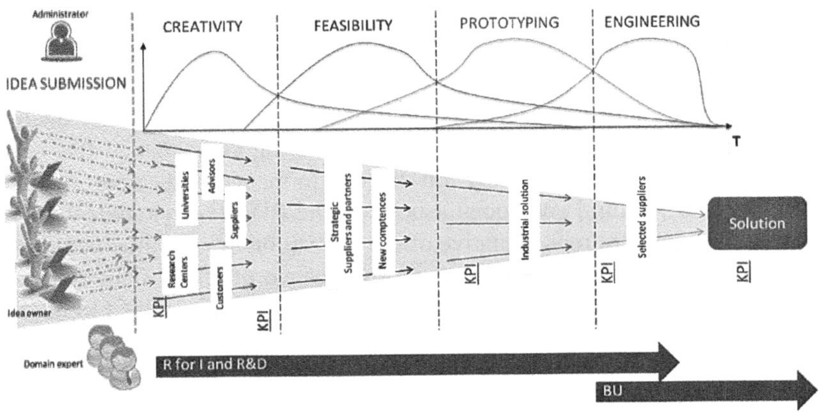

Figure 14.1. *Innovation waves*

The SMC observed the relevant aspects related to innovation projects management of Loccioni Group, and to KPIs collection connected to innovative potential of a team or of a single project.

14.3. Virtual innovation factory platform

Before starting the SMC, as end users, Loccioni collaborated with technical partners in order to detect technical problems, fix bugs and to indicate the "expected usefulness" of the platform components and functionality setup.

The VIF functionalities[4] that are tested during the preparation phase are the following:

– personal dashboard, where one can check all the personal notification, personal information about personal profile, and use the calendar;

– idea dashboard, where it is possible to insert a new idea specifying the issue, the title, the description, some tags, document in attachment and cover, comments, the choice of the members inserting the name or directly the skills connected to the idea, the visibility choosing between the VE, BE and the public (PU). Then, the end user can choose whether one wants to submit the idea to a domain expert who can evaluate the posted idea and launch an innovation project or simply save the idea in the draft panel, sharing it only with selected members;

– innovation project panel that shows the posted ideas submitted to domain experts and the launched innovation projects. When an innovation project is selected, it is possible to enter into the specific project page where the user can set the calendar, the Gantt and the project member; he can submit the document, insert comments and check the project status;

– observatory panel that enables to select the sources that could come from social networks, semantic Web crawler, and Web search;

– collaboration panel and chat tool with meeting rooms or forum and chat.

Several internal tests are conducted in order to improve the idea visibility. When we create the idea in the draft panel, one should share the idea wizard only with selected members, while when one submits an idea, he should share it with domain experts with specific competencies in the field of research of the proposed idea that should evaluate whether the idea is good to start an innovation project.

The role of domain expert is a very delicate aspect. In fact, the crucial argument of discussion is related to the choice of domain expert. *Should the expert be chosen by the idea owner or is there a list of domain experts that could evaluate the idea? In this case, when can one launch an innovation project? When just one expert gives positive feedback, is there a minimum number of positive evaluations?* In any case, the idea should be attached to a specific domain because the choice of domain experts may be more accurate. A deadline for the evaluation is necessary because usually the time elapsed

4 BIVEE deliverable D4.21 – virtual innovation factory prototype (v 1.0), see http://wordpress.bivee.eu/resources/publications/.

between posted idea and creativity wave start needs to be very short. With respect to VE management, once the idea is inserted, Loccioni prefers to decide with whom to collaborate in a VE in order to carry on the innovation project, considering the list of members chosen for their skills or for their degree and the list of selected comments related to the idea. The comments are not inserted only from chosen members but it should depend on the visibility that was set. The visibility level, which identifies the group of users that can visualize the idea after the submission, can be chosen coherently in this wave with the decided open innovation approach. The idea can be shared with the VE, the BE or the public. So one can choose, for example, to collaborate with a "BIVEE user" who proposed a valid solution for an idea. For this purpose, a VE can be created related to the single innovation project and we should set the visibility of the posted idea choosing from Loccioni BE, from other BEs or from a list of VEs.

The evaluation has shown that the role of an innovation administrator is required to check legal rights in order to accept an idea and to check the reliability of user profiles.

Finally, the innovation administrator should coordinate the evaluation of the ideas by domain experts, establishing reviewers, deadlines, etc. This evaluation should be obtained very quickly; therefore, this cannot be obtained only through a spontaneous candidacy of a domain expert but should be pushed by an administrator.

14.4. KPI selection and BIVEE platform prototype

One of the methodologies in BIVEE is to monitor and determine the current condition of an enterprise through values of a set of variables called KPIs.

Starting from a list of approximately 40 KPIs used in the first monitoring campaign, we proceeded with the definition of the formula for these KPIs[5].

For example, the KPI related to cost is equal to

$$\text{Cost KPI} = \text{material} + \text{labor} + \text{other} \qquad [14.1]$$

5 BIVEE deliverable D6.12 – analysis of user requirements, see http://wordpress.bivee.eu/resources/publications/.

In the original KPI list, many items were repeated for the activities of each wave. Thereby, the number of KPIs was increased from 41 to 126 (Figure 14.2(a)).

WAVES	ACTIVITIES				# of KPI
Creativity	Idea Generation	Idea Analysis	Initial Market Analysis	Final Assessment of the Idea	30
Feasibility	Resource Analysis	Feasibility Study	Proof of Concept		27
Prototyping	Resource Allocation	Design Prototype	Validate Prototype		32
Engineering	Acquire Resources	Build and Optimize	Launch Planning	Release	37
				TOT	126

a)

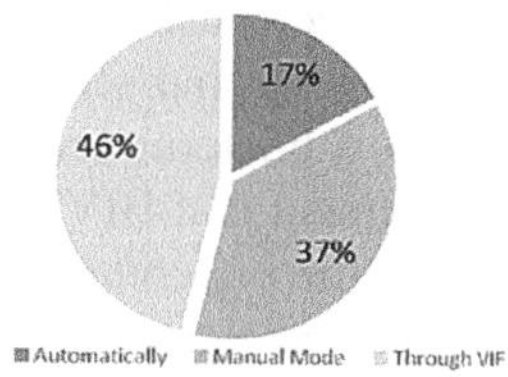

b)

Figure 14.2. *a) Total number of KPIs and b) types of KPIs collection*

After the KPI definition, the KPIs were categorized into three categories: automatically collected, manually collected and directly extracted from VIF data. As we can observe from Figure 14.2(b), there is a relevant number of KPIs that could be extracted by VIF or manually while there is a low number of KPIs obtained automatically updating these data from Loccioni's database. Then, the KPIs collected from VIF are useful because KPIs that are related to the creativity wave cannot be introduced automatically, but the whole procedure of innovation potential followed by Loccioni Group can be monitored since the application scenario is not related only to a single project.

As already explained Chapter 7 of this book, the BIVEE environment prototype[6] is composed of a monitoring platform that includes the RDH component and the data storage, the service platform and BIVEE portal that includes the front-end to the BIVEE platform and components. At the moment, through BIVEE portal in the monitoring platform, we are collecting manually or automatically the KPIs related to the projects monitored.

14.5. Conclusions

One of the most important aspects of the SMC is the continuous interaction between the platform and its users. In fact, having the BIVEE

6 BIVEE deliverable D6.32 – BIVEE environment prototype (v 2.0), see http://wordpress.bivee.eu/resources/publications/.

environment in action during the SMC, end users have been able to provide feedback about their experience related to both cultural and technical aspects. This valuable feedback has been addressed during the platform fixing and refinement activities to give to the BIVEE environment its final shape. This enables technical partners to work on detected issues and possible improvements and see the user reaction in a much faster manner. Therefore, there is a bilateral improvement related, on the one hand, to the cultural improvement that BIVEE framework provided to end users and, on the other hand, to the big enhancements introduced in BIVEE environment due to the feedbacks collected from end users and stakeholders.

From the end user's point of view, some internal procedures have to be reorganized in order to understand whether the new methodologies introduced by BIVEE platform improved innovation project management. First, we had to reorganize the way to create an internal "work order" and how to automatize some operations that were done manually before. The implementation was not a simple and short procedure, but meetings and formal steps have been organized before the complete adoption of the VIF that has been completed with the SMC.

14.6. Bibliography

[CHE 03] CHESBROUGH W., "The era of open innovation", *Sloan Management Review*, vol. 44, no. 3, pp. 35–41, 2003.

[DEP 12] DE PANFILIS S., MISSIKOFF M., "A knowledge-centric approach to virtual enterprise innovation", *Proceedings of the CEUR Workshop*, vol. 864, pp. 1–10, 2012.

[LAZ 10] LAZZAROTTI V., MANZINI R., PELLEGRINI L., "Open innovation models adopted in practice: an extensive study in Italy", *Measuring Business Excellence*, vol. 14, no. 4, pp. 11–23, 2010.

[RAN 13] RANGUS K, DRNOVŠEK M., DI MININ A., "Proclivity for open innovation: construct conceptualization and empirical validation", *Proceedings of the Acad Manage, Academy of Management Journal*, January 2013.

[SCH 11] SCHROLL A., MILD A., "Open innovation modes and the role of internal R&D: an empirical study on open innovation adoption in Europe", *European Journal of Innovation Management*, vol. 14, no. 4, pp. 475–495, 2011.

15

The AIDIMA Experience

This chapter describes the experience of AIDIMA during the BIVEE project and the activities performed regarding the definition of the pilot application cases, the configuration of the environment provided by business innovation in virtual enterprise environments (BIVEE) in the Production Space and the analysis of the impact produced in the furniture manufacturing sector represented by AIDIMA.

High-level descriptions of the validation scenarios, followed by an overview of the monitoring campaigns, are included in this chapter. Then, the activities related to the BIVEE setup of the furniture production environment for the monitoring campaigns and those issues that are detected during the preparation period are presented. To conclude, the analysis of the impact produced by BIVEE in the furniture manufacturing sector is explained in detail.

15.1. Introduction

The experience of BIVEE in the Production Space is represented by AIDIMA, which associates furniture manufacturing companies from Spain. The industrial scope represented by AIDIMA covers a wide range of small and medium-sized enterprises (SMEs) from the sector, being many of them traditional companies with basic technological level implemented in its infrastructures. The pilot application case selected for the validation of the BIVEE environment tries to represent a generic Virtual Enterprise (VE) which meets the features of many production networks of the Spanish furniture sector.

Chapter written by Fernando GIGANTE VALENCIA.

During the validation of BIVEE, it goes under two consecutive monitoring campaigns. The first monitoring campaign (FMC) monitored the state at an initial stage (AS-IS) of the end-user organizations. The second monitoring campaign (SMC) made an assessment of the use of BIVEE according to the delivered pilot application validation cases which are explained in more detail later in Chapter 15.2. Both monitoring campaigns collected valuable outcomes in order to achieve this analysis and validation. A comparative impact analysis for assessing the impact of BIVEE on the organizations is also presented.

The validation and evaluation methodology is based on the use of the tools provided by BIVEE after deploying them in the furniture environment. The comparison between the state of the art of this industrial environment and the new situation driven by the presence of BIVEE makes the measurement of the impact produced by the BIVEE environment possible, especially in those cultural aspects.

15.2. Validation scenarios

For the development of the production space scenarios, the current state of the production management activities was analyzed, identifying the most important processes and the main actors involved. Several meetings were arranged with external companies and customers of AIDIMA. The situation and difficulties of each company were evaluated in order to find the most interesting application cases according to the expected functionality of BIVEE. A first version of the application cases was released in order to be improved and refined later to perform the validation.

For the production improvement in the furniture industry, AIDIMA defined two validation cases. The first case was based on the production storyboard of one special item, while the second case was based on the furnishing of a kitchen. Regarding the great similarities of both application cases and considering the larger scope and possibilities of the first defined scenario, the effort was later focused on the improvement of the first application case. This scenario is based on the VE of Muebles Romero which represents a networked production approach, emerging but not yet widely implemented in the Spanish furniture industry. The scenario proposes an improved production network driven by the management of this network through the BIVEE environment.

The application scenario was defined on a VE so that it implies "cross-enterprise" processes with a set of heterogeneous companies involved. The selected scenario is not composed of only optimal cases; it has been selected in a realistic perspective. The pilot case also reflects possible problems that may arise during the value production processes, such as delays, faulty production, etc. This ensures that monitoring capabilities, support to exception handling and contingency planning functions of BIVEE could be extensively tested during the validation process.

The production of the selected product requires the participation of at least four or more external partners, including suppliers of raw materials, external producers for some product parts and external producers to perform special finishing operations. The number and names of these production partners depends on the configuration requested by the end customer.

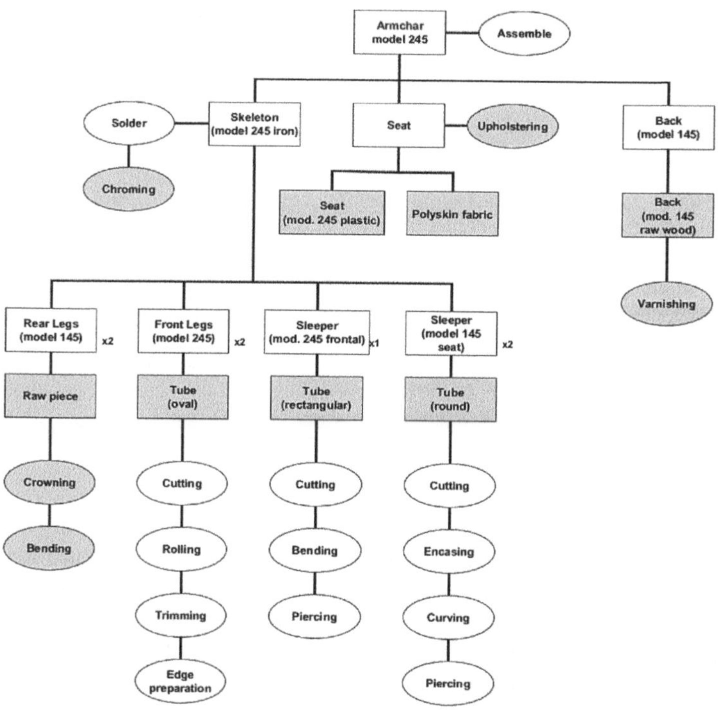

Figure 15.1. *Production breakdown of the armchair model 245 by Muebles Romero*

The composition of the armchair 245 is formed by some components shared by other furniture items from the catalogue of Muebles Romero. For this reason, the rear legs, the sleeper and the back of armchair 245 are annotated as "model 145" in the diagram above; these parts are used in both models but are internally labeled using the second name.

The boxes with shaded backgrounds represent the raw materials and semi-elaborated products supplied by members of the VE. Regarding the processes, those with transparent backgrounds represent the internal processes developed by Muebles Romero while those with shaded backgrounds represent the processes which are implemented by the other members of the VE.

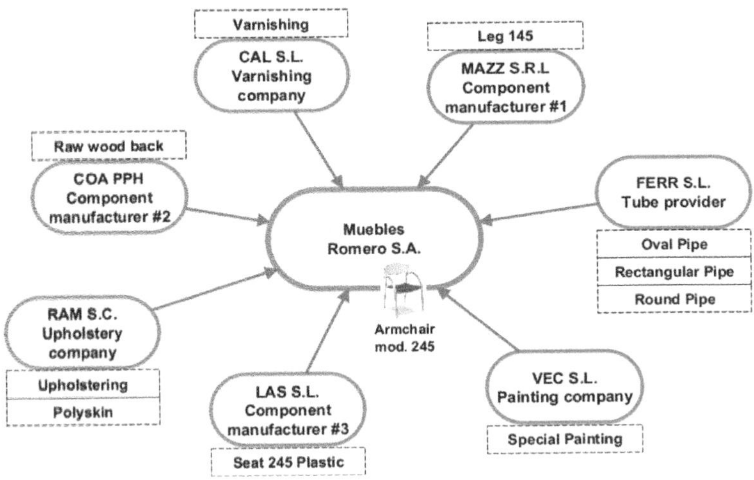

Figure 15.2. *Virtual Enterprise for the manufacturing of armchair model 245 produced by Muebles Romero*

As shown in Figure 15.2, despite being quite a simple furniture item, the production of this armchair involves the participation of different partners and there is a high dependency among the activities performed by each one. This figure represents the VE defined by Muebles Romero in order to produce the chair included in the proposed scenario. This network is mainly formed by traditional furniture SMEs.

The skeleton of this armchair is formed by a tube with added steel. The seat is made of varnished beech plywood and it is upholstered. The back is also made of varnished beech plywood. It has been designed only as an indoor furniture item and its design allows this furniture piece to be stackable.

During the armchair product life, the required operations have been performed through internal resources and external providers following different configurations. These different combinations could be considered as the specific instances of the production plan of this furniture item.

However, this is a very customizable product and that is the reason why this is one of the star products from the catalog. It is also for that reason that this model has been selected to replace the previous model. The customizable elements of this product are the color of the seat and the color of the back (both components are made of beech wood). Some external operations are required from external partners in order to produce this item, such as raw materials supply, external manufacturing of some components and special services considered as external operations.

The proposed scenario describes how the BIVEE environment is used to manage the collaborative production activities carried out in the VE. In order to summarize, a general description of the scenario including the most relevant issues is presented in Table 15.1.

Production Phase	Description of the scenario
Plan Phase	The scenario is based on the request of 300 chairs for the furnishing of two restaurants in a hospital complex located in Palma de Mallorca. Purchase managers from the hospital are interested in a customized version of the armchair 245 from the product catalog of Muebles Romero. After several meetings between the representatives of both organizations, the order is formalized. The order requested is later evaluated and some critical issues are detected. These are mainly the significant volume of the order and a tight delivery deadline.
Source Phase	The furniture company considers that the involvement of some external partner is necessary to perform some

	production activities in order to fulfill the order. The Product Breakdown is one of the key documents in order to manage the special configuration of the product and the internal and external operations to be performed. At this point, the mission control room (MCR) provided by BIVEE enables the definition of the whole production process, including a set of tentative production partners. The availability of indicators from these stakeholders enables a reliable simulation of the production processes inside the defined Virtual Enterprise allowing the evaluation of the production network.
Build Phase	The Product Breakdown is instantiated and the processes involved in the sales order have been submitted to BIVEE. During the production activities, some unexpected events arise. The production of the seat components finishes ahead of schedule while some delay in the manufacturing of the raw back components is reported by the corresponding partner. Due to this delay, the partner in charge of the varnishing operation is not able to start its activities on time. In the face of that problem, another partner is selected to finish that missing part of the order. At this point, BIVEE provides a whole vision of the production network including KPI monitoring to evaluate the performance of the Virtual Enterprise.
Delivery Phase	BIVEE provides a reliable estimation of the delivery time. Furthermore, the MCR enables the measurement of the customer satisfaction through the monitoring of those customer-oriented KPIs.

Table 15.1. *Summary of the key points addressed in the proposed furniture scenario*

15.3. Monitoring campaigns

The goal of a monitoring campaign in an enterprise is to determine the current condition of the enterprise through some variable values. Adopting key performance indicators (KPI) for monitoring is one of the commonly used approaches in organizations. KPIs are the critical measures that focus on organizational performance for the current and the future success of the organization [PAR 10].

On the one hand, during the first monitoring campaign (FMC), the AS-IS status of the companies without the BIVEE environment was examined. In light of the results of the FMC, the user requirements and the pilot validation scenarios were updated. Some new KPIs were introduced according to the FMC results and the input of the involved stakeholders.

The SMC, on the other hand, monitored the companies with the first deployable version of BIVEE. The outcomes of the SMC preparations in the furniture environment are presented in this chapter.

After the collection of measurements for the selected set of KPIs for a 4 month period in FMC, AIDIMA made an assessment analysis on top of those numeric values in order to look at the current status of the end users from a higher perspective. Each KPI relates to a business objective: reliability, velocity, adaptability, cost orientation, asset orientation, innovative potential, customer orientation and network orientation. The values determine how successful the company was with the specific objective. The number of KPIs selected for each profile in each use case also demonstrates how important business objectives are for the success of the VE.

From the AIDIMA assessment, it is reported that companies from furniture sector are very much interested in velocity and customer orientation profiles while they do not have much KPIs specific to the adaptability profile. This arises from the fact that the case presented by AIDIMA focuses on production lifecycle where customer demand is the driving force and production cycle time is an important performance factor [GIG 14].

Figure 15.3 shows some of the most relevant indicators identified for the measuring of the performance of the selected VE focused on furniture manufacturing.

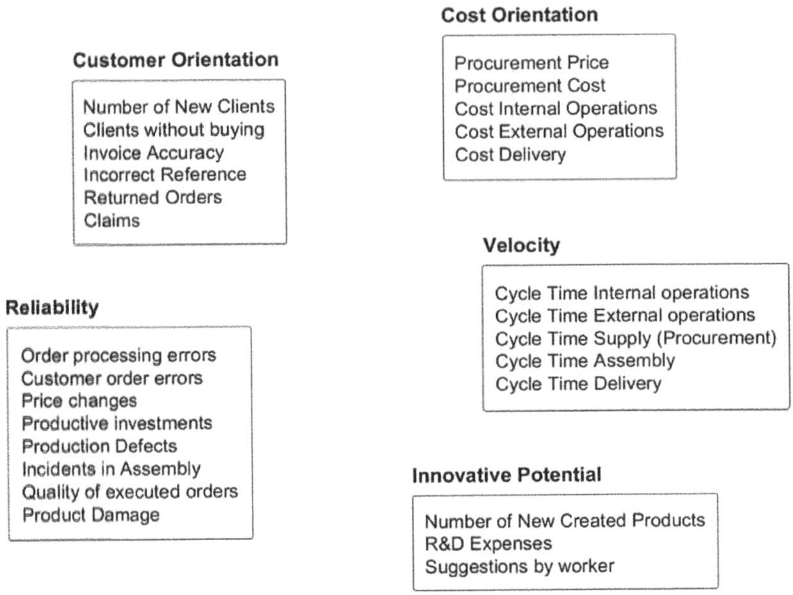

Figure 15.3. *Overview of KPIs identified for the furniture manufacturing Virtual Enterprise*

Some social aspects were also taken into account when analyzing the results. One way of categorization is a classification of the measured KPIs: whether it is internal or external to the specific enterprise. Another way of categorization is to make a classification whether a KPI can be considered as human centric or not.

During the collection and analysis of KPIs, tacit knowledge about the business of the associates of AIDIMA was gathered. This tacit knowledge can direct to some target areas where BIVEE applications can propose improvements, which are among the expectations of the SMC.

The length of SMC was extended until the end of the project in order to continue using the tools collecting more feedback from the end users. In this period, the list of KPIs was updated again including indicators related to the use of BIVEE and some internal operations performed by Muebles Romero. These last additions were done in order to perform a better validation of the selected scenarios; so this also considers these internal activities together with those key collaborative processes.

The list of KPIs was managed through an internal Excel file in order to keep track of which KPI values have been submitted to the BIVEE environment and how should they be introduced. This submission can be done inserting the values by hand through the Raw Data Handler (RDH) or automatically by using the ETL (Extract, Transform and Load regarding database processes) Plug-in under the Talend Open Studio[1] (TOS) environment.

In order to keep track of the status of the KPI collection, a spreadsheet was used to monitor the submission of the values of each indicator. This file was internally shared among those people involved in the monitoring campaigns and this allowed the introduction of comments in order to ease the KPI data collection process. The way the KPI values were collected from the different companies involved in the production network of Muebles Romero should also be noted. In this regard, the following cases can be defined:

– *automatic*: these values can be obtained directly from the Enterprise Resource Planning (ERP) of Muebles Romero and the corresponding VE members. As in the previous period of the SMC, the ERP developed by AIDIMA (GdP[2]) has been used to get these values if they were stored in the corresponding database. In this case, different ETL jobs were defined by using the BIVEE ETL Plug-in to retrieve the values from the database of GdP, the furniture ERP. Some examples of developed ETL jobs are : (1) new clients, (2) sales by product reference, (3) average value of processed orders, (4) newly created products, and (5) the number of processed orders;

– *semi-automatic*: this is formed by those indicators whose values cannot be directly obtained from the database, but they can be deduced by a manual processing of some value stored in the database and another value obtained manually. Some examples of semi-automatic KPIs are: (1) quality level of executed orders, (2) percentage of delayed deliveries, (3) invoice accuracy, and (4) percentage of returned orders;

– *manual*: these KPIs cannot be obtained directly or indirectly from the database of the ERP system. Hence, they should be obtained from documents. For the given list of KPIs, the referred documents have been mainly invoices, delivery notes and processed claims. Some examples of manual KPIs are: (1) cycle time of order evaluation/quotation, (2) cost of product definition, (3) production capacity and (4) cost of delivery.

1 http://www.talend.com/products/talend-open-studio/.

2 http://www.aidima.es/gdp/.

Figure 15.4 illustrates the percentage of KPIs which are retrieved in an automatic, manual and semi-automatic way. The left side of the figure represents this distribution, focused on the scope of Muebles Romero. On the right, the percentages are related to the members of the VE proposed in the pilot application case for the furniture production case. The intensity of the shadowed background represents the automation level for the collection of KPI values.

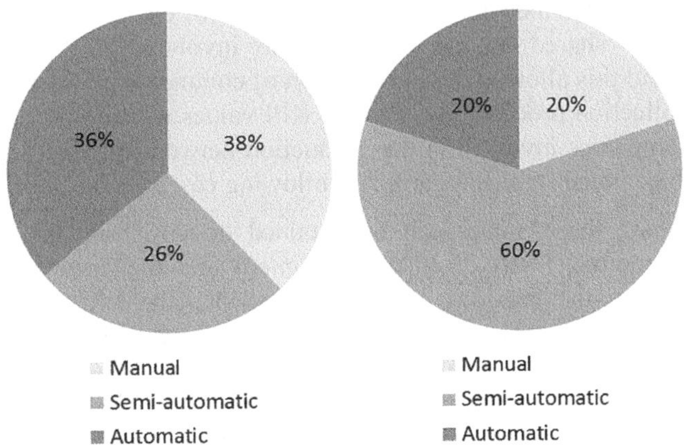

Figure 15.4. *Percentage of KPIs according to the collection mechanism; Muebles Romero (left) and other members of the Virtual Enterprise (right)*

Regarding the users, some distinction should be made according to the level of involvement in the project and the production activities. There are two kinds of users:

– *BIVEE-aware*: these users took some basics about the project and were able to manage those tools that are more oriented to technical people such as GdP (furniture ERP), the ETL Plug-in under TOS and the KPIOnto Editor. Furthermore, they have also used the RDH and the mission control room (MCR);

– *BIVEE-foreigners*: these users belong to the Production Space and most of them are from external companies associated with AIDIMA. These users are usually in charge of production activities and they have used the RDH partially to submit the test tool and KPI values, and mainly the MCR.

The RDH is the tool used to submit the KPI values which only cannot be calculated automatically. A new fact should per each KPI value should be created in the tool to register these values in the environment. This interface provided by BIVEE is useful but it is not user-friendly, especially if there are many KPI values to be submitted manually.

In this regard, another approach was considered in order to automate as much as possible the population of KPI values in BIVEE. Many companies are not able to provide values for its KPIs directly from an ERP or another management system. To address this issue, many KPI values should be introduced manually. This approach consists of using a spreadsheet to introduce these values and then defining an ETL job to submit them.

The spreadsheet was defined with a specific structure based on the dimensions required to submit the values properly. In the case of the indicators from the production space, the most relevant are those related to (1) VE level, (2) enterprise, (3) department, (4) subphase/subwave, and (5) product level. Furthermore, parameters such as time (month) and KPI value have been considered to fulfill the correct submission.

When creating the definition of this approach, the objective was to include the name (or identifier) of the KPI as another parameter in the spreadsheet, but this was not possible; therefore, the tRDHLoader – the component accountable for receiving the final value and submitting it to the BIVEE environment – could be only assigned to one specific KPI through the corresponding metadata. It is not possible to make this dynamically adaptive because each KPI is entirely different from all others in terms of data structure. Then, one different tRDHLoader per KPI was added in the ETL job.

However, it was possible to use one single tMap component – in charge of making the mapping between the input and the output parameters – in the ETL job, instead of using different mapping components per each indicator.

Technical partners provided some suggestions for this issue. They asserted that this could be addressed by using conditional sentences to discern the output tRDHLoader to be used depending on the KPI name.

After some evaluations, this last option was finally discarded. It was noted that there was no need to have the KPI names as parameters in the same spreadsheet because the quantity of values could make the management of this spreadsheet very confusing. As an alternative, the values of each KPI were introduced in different tabs, also maintaining a unique spreadsheet. In this way, each tRDHLoader is assigned to one different tab in order to read and submit the values of each specific indicator.

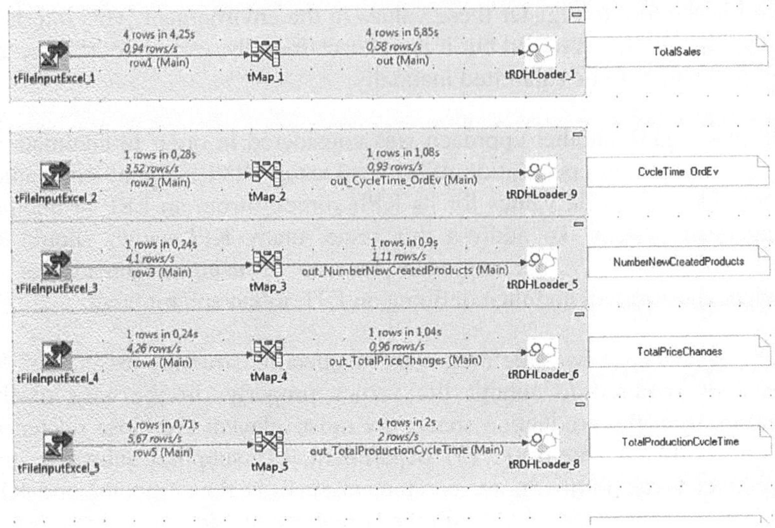

Figure 15.5. *Definition of unique ETL job to populate "manual" KPI values*

15.4. The BIVEE setup

In this section, we describe the key aspects to be considered during the introduction of the VE in the business ecosystem of the BIVEE environment. The issues addressed in this section are based on the BIVEE experience from AIDIMA and it can be extrapolated to any other production environment. The steps are explained in details in the following sections.

15.4.1. *Product definition*

This step involves the selection of products whose production is intended to be managed by BIVEE. Given the scope of BIVEE, it is recommended to select a product whose manufacturing involves a group of different production partners in order to get a valuable advantage of using the developed system.

One of the first steps to set up the VE is the definition of those partners involved in the manufacturing of the product selected. This was done by creating a Scope model through MCR modeler.

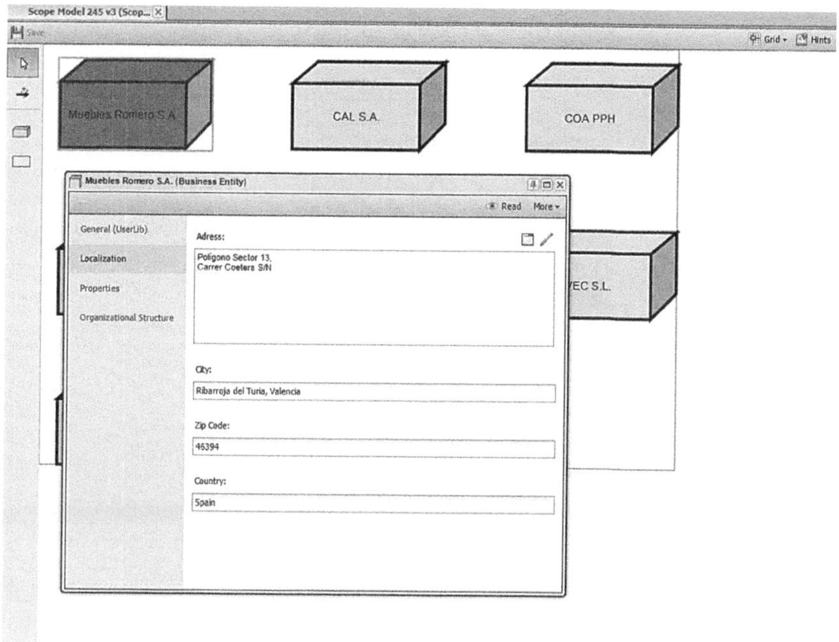

Figure 15.6. *Definition of partners involved in the production activities*

The diagram is formed by "business entities" which represent companies or specific parts of a company. Relationships between these entities can be defined. Furthermore, some properties for the given enterprises are introduced in the model.

The structure of the product is introduced in a product description diagram in order to define the different parts of the product and the materials involved. Figure 15.7 shows this product structure and the introduction of some relevant criteria in the properties panel. The most relevant custom features introduced are (1) description of the element, (2) type of element (i.e. component, part and raw material), (3) quantity of items required, (4) unit of measure and (5) cost of the given element.

Furthermore, a partner search can be performed in order to define the partner responsible for producing or providing each part of the product.

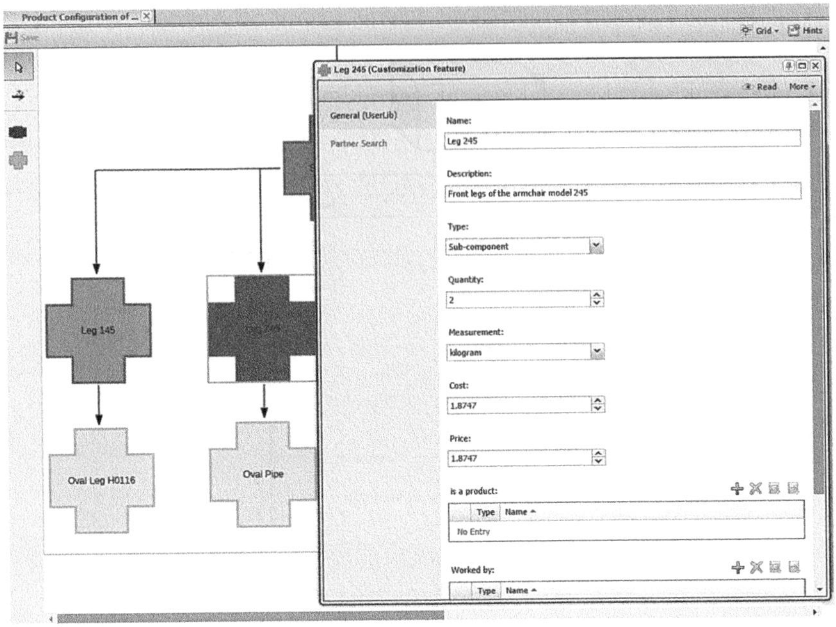

Figure 15.7. *Definition of the structure of armchair model 245 by Muebles Romero*

15.4.2. *Definition of processes*

Processes involved in the production of the selected product should be defined in the system. BIVEE provides the MCR tool to define the required processes and establish relationships between them. "The thread model" provided by the MCR allows the definition of collaborative processes through the use of swim lanes in the diagrams. This enables an improved vision of the dependencies among the activities carried out by the different production partners.

In the pilot application case, the process definition has not been too detailed for several reasons. On the one hand, efforts have been made in presenting a simple and intuitive scenario to ease the vision of the production environment and the produced impact on it.

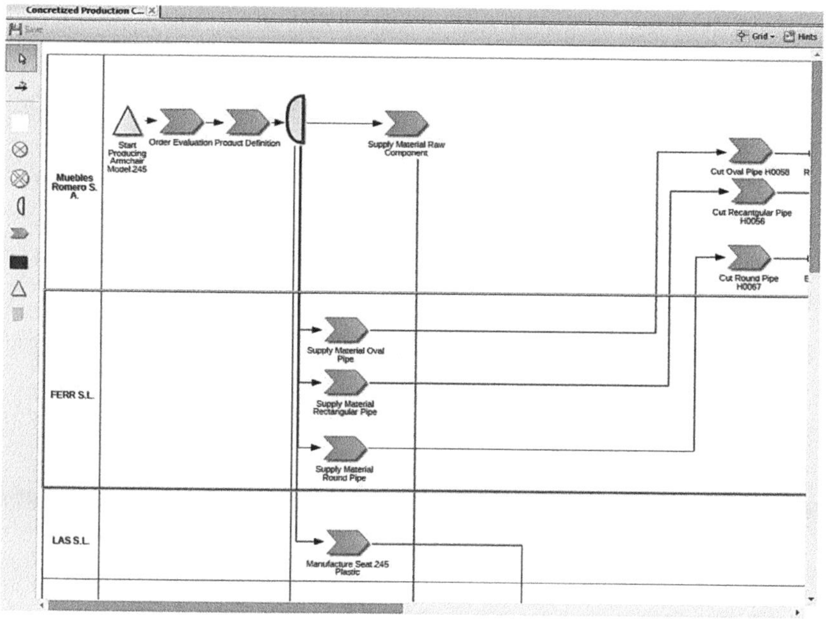

Figure 15.8. *Definition of the production processes for the manufacturing of armchair model 245 by Muebles Romero*

A high level of detail would be necessary in order to represent specific operations developed in the internal scope of each company. This can be done through the MCR, but this remains out of the scope of BIVEE which is not focused on the management of the internal operations but on those collaborative activities.

15.4.3. *KPI management*

One of the most critical points of the BIVEE setup is the one related to the KPIs in two main aspects: the selection of indicators and the measurement of its values.

Involved technicians from inside and outside the AIDIMA consider that, in a real production environment, companies should be able to define new indicators using the mechanisms provided by BIVEE. In this regard, the technical partners released the KPIOnto Editor, a user-friendly interface to edit the content of the KPI ontology.

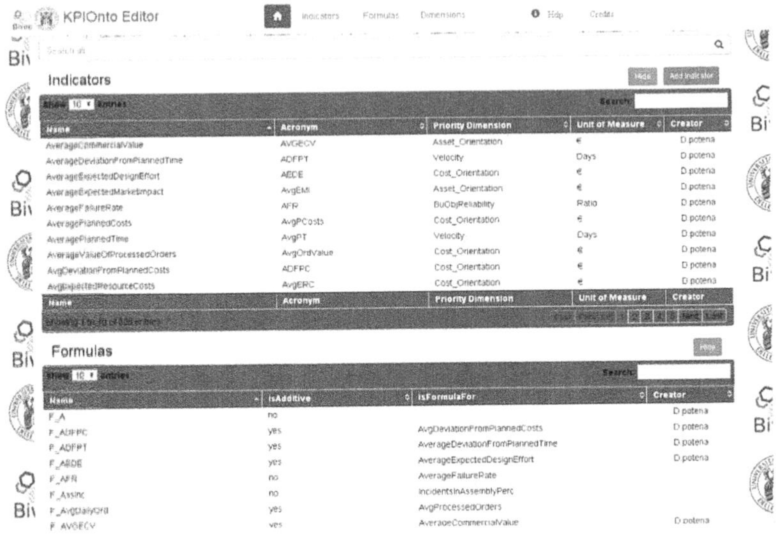

Figure 15.9. *Detail of the KPIOnto Editor to manage the KPI ontology*

As previously explained, one of the most important objectives of the SMC was to achieve a more automatic way to obtain the values of indicators and also an integration of these measurement mechanisms with the BIVEE environment. During the FMC, some values could be obtained directly from the interface of GdP: the selected furniture ERP. The way the value of these indicators has been retrieved has been explained in section 15.3.

For the automation and integration of the measurements, the BIVEE technical staff proposed the tool TOS to establish a bridge between the legacy system of the companies and BIVEE regarding the measuring of KPIs.

15.5. Encountered issues

This section describes the main roadblocks encountered during the adoption of the BIVEE tools in the production environment. It should be remarked that these problems are not related to the tools but the approach proposed by BIVEE during the monitoring campaigns and the validation of the environment. These are specific challenges to be addressed and they should be considered when using BIVEE in a given industrial sector.

15.5.1. *Furniture production cycles*

The furniture industry is characterized by unexpected production cycles depending on the ongoing sales and the expected sales. The length of the SMC was not long enough to reflect the sales variability and the production changes. Despite this, the SMC experience was anyway feasible and useful even though this was a short time frame. This aspect affects the SMC; therefore, it covers a different time frame than the one covered by the FMC. In the furniture case, the selected furniture manufacturer Muebles Romero manages a different sales forecast for the first half of this year than the one of the last few months of the previous year. The product selected for the FMC will be probably being produced during the course of the SMC, but in a lower production volume. To solve this during the project period, a referential product was selected. This was done with the close support of the production responsible staff of Muebles Romero.

The selected referential product shares almost the same production process than the one managed during the FMC. Furthermore, they share the main production partners allowing a homogeneous view of both monitoring campaigns in order to perform a comparative analysis of the results of both campaigns.

15.5.2. *Data sharing*

One of the most outstanding obstacles for the SMC was detected during the course of the FMC. This was the privacy of information, due to fear of sharing sensible KPI information even among the members of the production network.

In the pattern commonly used in the current production activities, relationships are based on the client-supplier model which is not a collaborative approach. Most companies are reluctant to share their data with other production partners. Among others, they consider that their customers could increase the level of demand if they have too much information about their processes.

This issue has been progressively solved introducing this new approach to the companies, explaining them the advantages for all the members in a real collaborative production environment and participating in other projects related to this emerging approach. However, this is an important cultural aspect to be considered during the adoption of any collaborative system like BIVEE in any industrial environment.

15.5.3. *Low technology SMEs*

Not all the members of the selected VE have tools available to obtain their KPI values automatically. The production network proposed in the pilot case of Muebles Romero reflects a real collaborative production environment with heterogeneous SMEs, which perform its activities managing different levels of technology.

The objective of the monitoring campaigns was to not force any company to start using specific technologies to carry out its internal activities. Alternatively, the objective is to provide tools to start managing its collaborative processes without causing, in principle, major changes in its current internal management. For those low-tech SMEs, the KPI values should be introduced manually by using the RDH interface provided by BIVEE. Only some certain values can be obtained from the ERP of Muebles Romero.

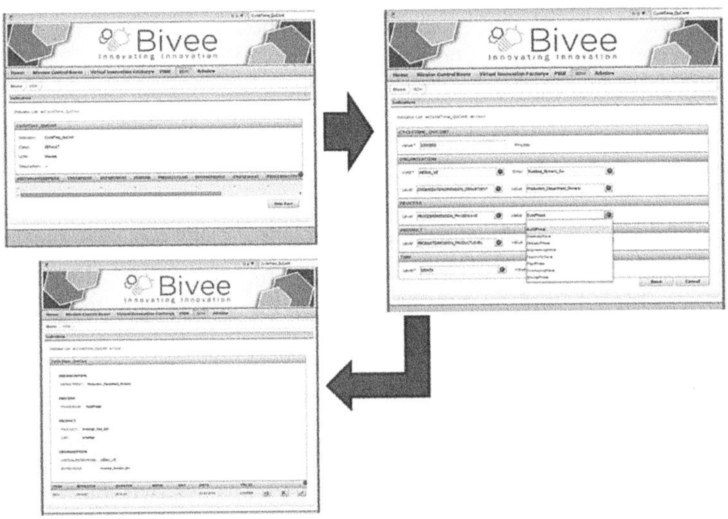

Figure 15.10. *Workflow of the introduction of new KPI values as facts using the RDH provided by BIVEE*

15.5.4. *Process modeling*

Companies are usually aware of the internal processes performed inside their facilities, but they do not usually have a complete picture of their complex or collaborative processes. It is important to pay attention to information exchange; therefore, in any collaborative environment, better communication mechanisms usually result in improved processes.

It is, however, difficult for furniture SMEs to dedicate resources to manage the tools provided by BIVEE because they usually require some previous skills. In that respect, AIDIMA has supported this task considering the current situation of many furniture companies.

Furthermore, during the SMC, some specific values for processes should be introduced through the MCR. Some problems were found in collecting some KPI values from companies involved in the VE of Muebles Romero. These are mainly values related to procurement cycle times and delivery terms. Some companies find the providing of that information hard because this could imply to acquire commitments under not well-known conditions. This point was explained to the involved companies arguing that, in principle, the provided information is required to start the checking and validation of BIVEE.

15.6. Improvements in the BIVEE environment

BIVEE tools have been gradually improved according to the feedback retrieved from users. In this regard, the Bitbucket[3] issue tracker has been managed in the production environment, especially for those issues that are related to the MCR. This has been a very useful tool to keep track of the progress in problem solving and those relevant suggestions proposed by users in order to improve the system. Besides this, many issues have been also managed by e-mail, remote calls and partner meetings.

Regarding the MCR, different issues, such as the operation of the model browser, the response to user actions, the management of panels and objects and the handling of exceptions, were improved. Some drawing aspects were solved, providing a high level of customization of the graphical interface. The theming package allows the customization of the working area for the users by shrinking, expanding or collapsing the different panels.

The second release of the MCR offers a renovated interface which provides a more integrated aspect of the tool in the BIVEE environment. Furthermore, user guidelines were also released. The user guide presents

3 https://bitbucket.org/.

clearer and better organized content to learn the use of the MCR. The document also explains the type of diagrams and basics about the tools in more detail.

The Spanish translation of the MCR interface was also an important issue addressed. The availability of having the texts of the MCR interface translated into Spanish was welcomed by users; therefore, many users from the Spanish furniture sector did not feel too comfortable managing such a technical tool fully in English. This enabled the involvement of more users in the testing and validation activities of the MCR, especially those from companies outside the AIDIMA.

The functionality of the RDH was also improved in some relevant aspects. The most important aspect was the new search capability. This new feature allows filtering of indicators by name according to the text introduced in the corresponding column. This speeds up the location of a given indicator in order to examine its values or submit new values to the system. Loading times were improved, especially in the KPI list. With this change, the navigation is more comfortable for the user. The browsing of KPI values and the submission of new facts were more user-friendly due to these enhancements.

Furthermore, the single sign-on (SSO) meant a substantial global improvement for the user experience. Users can manage only one single username and password to access the different tools of BIVEE. The BIVEE environment achieved a higher level of integration providing a unique sign-on point to access the different BIVEE tools.

15.7. BIVEE cultural improvement

The analysis of the impact produced by BIVEE in the furniture industrial sector cannot be measured only by comparing KPI values before and after the adoption of BIVEE. As explained before in this chapter, the short length of the monitoring campaigns did not allow observing significant changes in the KPI values. In addition, these campaigns covered different periods of the year. Considering that the production activities follow a cyclic approach, it is difficult to determine meaningful conclusions only from the comparison of KPI values. Besides this, production activities are often influenced by many factors. This hinders to determine whether any change in the KPI values has been produced because of the use of BIVEE or not.

To collect feedback regarding the cultural aspects influenced by BIVEE, a survey was submitted to associated companies and users. The following sections describe the most relevant cultural aspects affected by the use of BIVEE in the furniture manufacturing environment.

15.7.1. *Collaborative approach*

The Spanish furniture sector is mainly formed of traditional SMEs. These companies usually do its work following a client-supplier business model. This traditional strategy is quite different from the collaborative approach which is inherently proposed by BIVEE.

A collaborative approach aims to involve the different members of the production network in the achievement of one common objective. This is done by increasing the dialogue between the partners involved and increasing the flexibility of these partners. This approach is not focused on the improvement of the activities performed by one single member of the network but the improvement of the production activities as a whole.

The demands of the market often push companies to adopt a collaborative production approach in order to fulfill some specific orders. This aspect was represented in the scenario proposed for the Production Space.

BIVEE links ideas from the Innovation Space to processes in the Production Space in a way that any innovative proposal can be taken into account in order to improve a given production process. BIVEE provides the bridge between both spaces and provides mechanisms to define improvement proposals to processes through the MCR.

15.7.2. *Information sharing*

As mentioned before, most traditional companies from the furniture sector are reluctant to share some KPI information with other members of the VE.

Focusing on the proposed scenario, it should be emphasized that many companies of the production network did not have any kind of direct business relationship between them. The members of the selected VE only have in common their relationship with Muebles Romero, which is a common production partner for all of them. However, they did not have a reciprocal knowledge of each other in terms of a direct business relationship.

The availability of information about the different members of the network implies an important opportunity to improve the business of all these companies so that all the members have data available in order to evaluate the performance of the network.

This issue has been addressed for the production network defined in the Production Space scenario but, as commented previously in this chapter, this is an important challenge to be considered when adopting this approach in any other traditional industry.

The functionalities provided by BIVEE are especially useful to convince traditional companies of the benefits of sharing KPI information with other production partners in order to achieve a local and global improvement.

BIVEE changes the perception of the whole process by the manufacturers; therefore, it provides a global vision focusing on a given production target. With the BIVEE approach, companies feel part of a global network and they become aware of its importance and responsibility in this network to meet a common objective.

As such, the administrator of the VE acquires a new perception of the objective through a whole vision of the processes and partners involved. In particular, this provides a significant improvement when the objective involves many external partners, which is the main scope of BIVEE.

15.7.3. *Process management*

Since the beginning, BIVEE has proposed a production management based on processes and this has been especially reflected in the MCR.

BIVEE provides functionalities to properly define the different production processes through the MCR. This tool allows a high level of detail for processes, but the most important is the capability of managing those external processes undertaken by the different members of the VE. This functionality is not provided by the local ERP and it is provided by BIVEE.

In addition, the MCR provides the following additional features to support the process definition and management: (1) definition of dependencies between the activities performed by the different partners involved, (2) association between process and company (partner member of the VE), (3) association between process and person (responsible) in charge of the fulfillment of the process, (4) attachment of KPIs to given processes and (5) relationships between model diagrams (already defined) allowing a hierarchically structured view of the processes.

In addition, the use of the MCR implies a reduction in the time required to set up the production network. This also means a reduction in the cost associated with the corresponding administrative tasks.

15.7.4. *Detection of problems and opportunities*

The global vision about the processes carried out in the production network enables the detection of improvements. The relationships between processes and the association of these processes to specific partners allow for the detection of bottlenecks and problems during the production activities.

The MCR is able to detect an inefficient consumption of resources. The KPI monitoring functionality provided by the MCR informs the critical values according to the thresholds defined by the production manager for each indicator. The dynamic selection of partners which can be done after some unexpected problems, such as delays, errors in the order or product damage during the manufacturing activities, allows an automatic calculation of the time of completion and delivery. The MCR enables the measurement of the performance of the production network and it can be used as a support for the decision-making.

Hence, the response capability to face unexpected problems in production is significantly improved due to the global picture provided by the MCR.

In this regard, it is worth noting the impact produced in the customer, which is the key element in the furniture sector. The delivery phase is the most appropriate step to evaluate and analyze this business objective: the customer orientation. The following issues have a special influence in the customer satisfaction; furthermore, some of them have been already addressed directly or indirectly by BIVEE in the previous phases: (1) facilities to formally define a customized version of the product with the corresponding processes and dependencies between the partners involved; (2) BIVEE allows a more reliable determination of the delivery time to the final consumer; (3) capacity to monitor the customer satisfaction through those customer-oriented indicators, such as the number of clients, the percentage of orders returned, the percentage of claims received or the percentage of incorrect references; (4) feedback from the innovation space may potentially improve aspects about the satisfaction of the customer in terms of design, sustainability, quality of materials/components, final selling price, etc.; and (5) management of claims: company should manage some information about the claims received in order to populate the related indicators.

15.7.5. *KPIs management*

Finally, the management of indicators is also a critical factor regarding the cultural approach caused by BIVEE in the production space. Traditional manufacturers are not familiar with the KPIs management. In fact, in many cases, the only indicators managed by many companies in the furniture sector are those related to sales, such as the number of cancelled invoices, the percentage of wrong bids sent, and the ratio between the total number of sales and the expected sales.

The use of indicators strictly related to production is not widely implemented in the furniture sector; therefore, many of these traditional companies do not have mechanisms to measure them. Independently of these measurements, BIVEE means an inherent improvement in this area. This improvement is based on the management and monitoring of KPIs for those activities performed by all the members of the production network.

15.8. Conclusions

BIVEE proposes mechanisms to define a bridge between the innovations produced from the idea generation and the production activities that are improved by the innovation proposals. BIVEE also reduces the required effort in terms of resources to achieve this communication between the innovation space and the production space. This is especially relevant taking into account that these two spaces usually remain too far apart in the real business environments.

Furthermore, BIVEE proposes a change in the working methods of those companies that are not aware of the advantages of the collaborative manufacturing approach, the sharing of information and the KPIs management in a production network environment.

For AIDIMA, the activities performed during the monitoring campaigns have been an opportunity to increase the knowledge about what and how the SMEs are doing their production activities. This has also allowed the detection of its main problems and concerns.

The collaborative approach can also be considered as new and innovative to many traditional furniture manufacturers. However, the wide adoption of this approach in such a traditional sector, as the furniture one, requires time and effort. During the monitoring campaigns, effort has been dedicated to let the companies know the benefits of following this approach and illustrate through the use of BIVEE.

Process modeling immediately brings an improvement for the involved companies because these need to make a self-analysis and an evaluation of its activities in order to perform a proper process definition. This is especially important so that many traditional companies do not usually make this analysis. BIVEE also provides mechanisms to perform a continuous measurement and monitoring of the ongoing processes. Most of the companies only monitor their internal processes, ignoring the details of the work done by their production partners. With BIVEE, companies have new tools available, which lead to important changes in the managing methods. The availability of information about partners enables the companies to take actions for addressing deviations achieving the proposed production objectives.

15.9. Bibliography

[PAR 10] PARMENTER D., *Key Performance Indicators (KPI): Developing, Implementing, and Using Winning KPIs*, John Wiley & Sons Inc, New Jersey, 2010.

[GIG 14] GIGANTE F., MÁÑEZ R., SINACI A. *et al.*, "Preparation of BIVEE validation activities in furniture industry", *New Generation Enterprise and Business Innovation Systems – NGEBIS*, 2014.

Conclusions

The objective of this book has been that of providing the essence of the results of the European project BIVEE to a large audience. We hope that it will contribute to improving the innovation potential of the European SMEs and, more generally, to improving the awareness of all the people who contribute to any kind of innovation initiative. There is a growing understanding that innovation is one of the main factors that will guarantee for the European production system a solid way of regaining competitiveness against the most dynamic economies of the world, such as USA, China and India. And that innovation is able to guarantee a long-term economic success, rather than other forms of competitive strategies (e.g. the price war) that, if successful, may only bring a temporary, fragile competitive edge.

The book reported on the key points that have characterized the BIVEE project. Primarily, the idea is that innovation is essentially a *knowledge-centered* activity. In fact, when an enterprise starts an innovation venture, it needs a substantial set of preexisting competencies and a solid knowledge base, on which the challenging work of innovators needs to rely. Competences and knowledge should not be only on technologies and the specific application domain, but they need to span on a much larger territory. In fact, to achieve a successful innovation, it is necessary to cover and interconnect knowledge from many different domains: from enterprise organization to economics, from marketing to law. Thus, innovation requires us to produce additional knowledge, targeted to proceed toward a finding and solution that did not exist before. But, a knowledge-centric approach is also important for the optima management of an innovation project, from the

---

Conclusion written by Michele MISSIKOFF.

initial formation of the consortium, to select a set of partners with complementary profiles, to the day-by-day monitoring and management of the activities, milestones and costs. The great variety of knowledge to be managed in an efficient and effective manner requires the adoption of advanced semantic technologies; this has been a strategic choice in implementing the BIVEE platform.

The second key point concerns the focus of the project on *SMEs*. SMEs represent 99% of European industries, but when talking about innovation they cover a minor fraction of the innovation projects carried out. This is due to a number of causes, among which the high fragmentation, difficulty in establishing a smooth information flow with the potential partners, scarce practice with digital technologies and in particular *cooperation platforms* and last but not least the lack of the critical mass. This point is clearly addressed in the book, where the illustrated solutions are conceived to be adopted by virtual enterprises, specifically networked SMEs, pushing on open collaborative innovation. Here, one of the key challenges has been the problem of the large variety of information systems and data management solutions (some of them as simple as an Excel file) used in the various SMEs. Here, the project concentrated on an effective solution of *distributed data harvesting* and *semantic reconciliation*.

Another key achievement of BIVEE has been the conception of a framework, the business innovation reference framework (*BIRF*), aiming at proposing an engineering approach to innovation capable of preserving, and even supporting, the creative thinking that is the basic fuel of innovation. The BIRF is founded on the BIVEE methodology that in turn is based on the idea of four innovation waves: creativity, feasibility, prototyping and engineering. The metaphor of the wave intends to convey the notion of a propulsive power, a loosely structured organization of the activities, including fuzzy boundaries between the different waves that follow one another, and the extreme flexibility of a liquid substance. All the rest of the BIRF rotates around the four BIVEE waves, such as the document templates (and the corresponding document ontology: *DocOnto*), the monitoring systems and, primarily, the idea of the *virtual innovation factory* (VIF). The latter exploits the knowledge-centered approach of BIVEE, where innovation is seen as a set of dynamically organized activities having the objective of producing an innovative solution in the form of a "knowledge artifact". The production of this knowledge artifact is achieved by a knowledge factory, referred to as the VIF. A VIF is "virtual" for two main reasons, first since it has no predefined boundaries and organization, its components and the teams working for it will vary in time, depending on the specific wave, the addressed problems (sometimes unexpected), the available budget, the active "external" cooperation, etc. Thus, it is virtual in its final outcome: the "intangible" knowledge artifact. But, the overall production paradigm is similar to that of manufacturing production (but with a number of important differences): you start with an objective (but less clear), you acquire the necessary raw material (but of an intangible nature: the relevant preexisting knowledge), you refine the acquired material (filtering, selecting, connecting and organizing knowledge items), adding also new knowledge items, eventually producing the final knowledge artifact, ready to be transferred into production.

A vast literature clearly indicates that creative ideas represent only the spark that starts an innovation project. Thus, it is necessary to provide resources, to engage talented people, to spend time and money and to unleash enthusiasm. All those factors are mobilized in high-risk settings, where the probability of failure is lofty. For this reason, another key point is represented by a systematic and reliable monitoring framework, implemented by using a method and tool specifically conceived for the purpose. BIVEE monitoring framework is based on a system of *key performance indicators (KPIs)*, carefully selected, but supported by an open tool that allows the users to selectively adopt them and specify additional, custom KPIs when needed. Also, here the knowledge-centric approach is highly valuable, since the KPI system is supported by a dedicated ontology: PKI-*Onto*.

As anticipated, innovation cannot be carried out in isolation with respect to the rest of the world: it requires a constant cross-boundaries knowledge flow. The start is characterized by the acquisition of a first base of knowledge that includes, besides the knowledge on the specific target (product, process, service, etc), knowledge on the business context, market, competitors, etc. But, the knowledge should flow in two directions, not only "outside-in", with an intake flow, but also "inside-out", accepting external contributions. The latter flow is essential to maintain a constant and fruitful knowledge exchange with partners, research centers, clients and stakeholders, who may be of great help in several phases (waves). Such an *open innovation* paradigm needs to be implemented with a great care of the protection of intellectual property. BIVEE has been well aware that overprotection risks isolating the innovation process in an ivory tower with reduced probability of success, but on the other hand, full transparency may expose leakages, jeopardizing the competitive edge that original solutions bring.

A fruitful information and knowledge flow must also be established with the production space that represents the final destination toward which the innovative solutions will be conveyed. BIVEE is well aware of this and another central part of the BIVEE framework is represented by the notion of *value production space*, i.e. the production context where the virtual enterprise operates, from which large part of the innovation demand comes and where the innovation solutions need to be implemented. Here lies another critical issue: the capacity of the production organization to adopt new solutions and implement the changes that the acquired innovation requires. The specific part connected to the change management has not been systematically addressed within the BIVEE project and is in our agenda for a future project.

In conclusion, we are convinced that innovation can no longer be treated as an empirical discipline, and there is a compelling necessity to progressively introduce an engineering approach: BIVEE demonstrated that this can be successfully achieved. In the future, we intend to address the four main issues that emerged during the project: (1) promote educational programs to progressively build a diffused culture of innovation and, more specifically, the special competencies indicated in this book, including the role of innovation manager (see the e-Leadership proposal of the European Competency Framework[1]); (2) extend the BIVEE platform with the missing section necessary to guarantee a safe management of the open innovation approach by suitably protecting the intellectual properties; (3) systematically address the problems connected to the introduction of an innovation in a value production organization, i.e. a change management framework capable of supporting the introduction of innovations without reducing the production capability of the organization and; (4) but the very last point is the possibility of transforming the BIVEE prototype into a commercial, industrial strength product, and spread it among the European SMEs, to achieve the impact that the valuable outcome of BIVEE deserves.

1 www.ecompetnecies.eu.

List of Authors

Pierluigi ASSOGNA
Theorematica
Rome
Italy

Francisco CALLE MORENO
Atos
Madrid
Spain

Massimo CANDUCCI
Engineering Ingegneria
Informatica S.p.A.
Turin
Italy

Cristina CRISTALLI
Loccioni Group
Jesi
Italy

Claudia DIAMANTINI
Dipartimento di Ingegneria
dell'Informazione
Università Politecnica delle Marche
Ancona
Italy

Asuman DOGAC
SRDC – Software Research,
Development and Consultancy Ltd.
METU Technopolis
Ankara
Turkey

Nesat EFENDIOGLU
BOC
Vienna
Austria

Fernando GIGANTE VALENCIA
AIDIMA
Valencia
Spain

Mauro ISAJA
Engineering Ingegneria
Informatica S.p.A.
Rome
Italy

Daniela ISIDORI
Loccioni Group
Jesi
Italy

Benjamin KNOKE
BIBA – Bremer Institut für Produktion und Logistik GmbH
University of Bremen
Germany

Neil MAIDEN
City University
London
UK

Michele MISSIKOFF
National Research Council,
Institute of Sciences and Technologies of Cognition CNR-ISTC
Rome
Italy

Anil PACACI
SRDC – Software Research, Development and Consultancy Ltd.
METU Technopolis
Ankara
Turkey

Domenico POTENA
Dipartimento di Ingegneria dell'Informazione
Università Politecnica delle Marche
Ancona
Italy

Ali Anil SINACI
SRDC – Software Research, Development and Consultancy Ltd.
METU Technopolis
Ankara
Turkey

Fabrizio SMITH
National Research Council,
Institute of Systems Analysis and Informatics CNR-IASI
Rome
Italy

Emanuele STORTI
Dipartimento di Ingegneria dell'Informazione
Università Politecnica delle Marche
Ancona
Italy

Francesco TAGLINO
National Research Council,
Institute of Systems Analysis and Informatics CNR-IASI
Rome
Italy

Isabella TERZONI
Loccioni Group
Jesi
Italy

Wilfrid UTZ
BOC
Vienna
Austria

Robert WOITSCH
BOC
Vienna
Austria

Index

G, I, J, K

L, M, N, O

P, Q

R, S

T, U, V, W

Other titles from

ISTE

in

Systems and Industrial Engineering

2015

BOULANGER Jean-Louis
CENELEC 50128 and IEC 62279 Standards

BRIFFAUT Jean-Pierre
E-Enabled Operations Management

2014

CHETTO Maryline
Real-time Systems Scheduling
Volume 1 – Fundamentals
Volume 2 – Focuses

DAVIM J. Paulo
Machinability of Advanced Materials

ESTAMPE Dominique
Supply Chain Performance and Evaluation Models

FAVRE Bernard
Introduction to Sustainable Transports

MICOUIN Patrice
Model Based Systems Engineering: Fundamentals and Methods

MILLOT Patrick
Designing Human–Machine Cooperation Systems

NI Zhenjiang, PACORET Céline, BENOSMAN Ryad, REGNIER Stéphane
Haptic Feedback Teleoperation of Optical Tweezers

OUSTALOUP Alain
Diversity andNon-integer Differentiation for System Dynamics

REZG Nidhal, DELLAGI Sofien, KHATAD Abdelhakim
Joint Optimization of Maintenance and Production Policies

STEFANOIU Dan, BORNE Pierre, POPESCU Dumitru, FILIP Florin Gh., EL KAMEL Abdelkader
Optimization in Engineering Sciences: Metaheuristics, Stochastic Methods andDecision Support

2013

ALAZARD Daniel
Reverse Engineering in Control Design

ARIOUI Hichem, NEHAOUA Lamri
Driving Simulation

CHADLI Mohammed, COPPIER Hervé
Command-control for Real-time Systems

DAAFOUZ Jamal, TARBOURIECH Sophie, SIGALOTTI Mario
Hybrid Systems with Constraints

FEYEL Philippe
Loop-shaping Robust Control

FLAUS Jean-Marie
Risk Analysis: Socio-technical and Industrial Systems

FRIBOURG Laurent, SOULAT Romain
Control of Switching Systems by Invariance Analysis: Application to Power Electronics

GRUNN Emmanuel, PHAM Anh Tuan
Modeling of Complex Systems: Application to Aeronautical Dynamics

HABIB Maki K., DAVIM J. Paulo
Interdisciplinary Mechatronics: Engineering Science and Research Development

HAMMADI Slim, KSOURI Mekki
Multimodal Transport Systems

JARBOUI Bassem, SIARRY Patrick, TEGHEM Jacques
Metaheuristics for Production Scheduling

KIRILLOV Oleg N., PELINOVSKY Dmitry E.
Nonlinear Physical Systems

LE Vu Tuan Hieu, STOICA Cristina, ALAMO Teodoro, CAMACHO Eduardo F., DUMUR Didier
Zonotopes: From Guaranteed State-estimation to Control

MACHADO Carolina, DAVIM J. Paulo
Management and Engineering Innovation

MORANA Joëlle
Sustainable Supply Chain Management

SANDOU Guillaume
Metaheuristic Optimization for the Design of Automatic Control Laws

STOICAN Florin, OLARU Sorin
Set-theoretic Fault Detection in Multisensor Systems

2012

AÏT-KADI Daoud, CHOUINARD Marc, MARCOTTE Suzanne, RIOPEL Diane
Sustainable Reverse Logistics Network: Engineering and Management

BORNE Pierre, POPESCU Dumitru, FILIP Florin G., STEFANOIU Dan
Optimization in Engineering Sciences: Exact Methods

CHADLI Mohammed, BORNE Pierre
Multiple Models Approach in Automation: Takagi-Sugeno Fuzzy Systems

DAVIMJ.Paulo
Lasers in Manufacturing

DECLERCK Philippe
Discrete Event Systems in Dioid Algebra and Conventional Algebra

DOUMIATI Moustapha, CHARARA Ali, VICTORINO Alessandro, LECHNER Daniel
Vehicle Dynamics Estimation using Kalman Filtering: Experimental Validation

HAMMADI Slim, KSOURI Mekki
Advanced Mobility and Transport Engineering

MAILLARD Pierre
Competitive Quality Strategies

MATTA Nada, VANDENBOOMGAERDE Yves, ARLAT Jean
Supervision and Safety of Complex Systems

POLER Raul *et al.*
Intelligent Non-hierarchical Manufacturing Networks

YALAOUI Alice, CHEHADE Hicham, YALAOUI Farouk, AMODEO Lionel
Optimization of Logistics

ZELM Martin *et al.*
Enterprise Interoperability –I-EASA12 Proceedings

2011

CANTOT Pascal, LUZEAUX Dominique
Simulation and Modeling of Systems of Systems

DAVIM J. Paulo
Mechatronics

DAVIM J. Paulo
Wood Machining

KOLSKI Christophe
Human–Computer Interactions in Transport

LUZEAUX Dominique, RUAULT Jean-René, WIPPLER Jean-Luc
Complex Systems and Systems of Systems Engineering

ZELM Martin, *et al.*
Enterprise Interoperability: IWEI2011 Proceedings

2010

BOTTA-GENOULAZ Valérie, CAMPAGNE Jean-Pierre, LLERENA Daniel, PELLEGRIN Claude
Supply Chain Performance / Collaboration, Alignement and Coordination

BOURLÈS Henri, GODFREY K.C. Kwan
Linear Systems

BOURRIÈRES Jean-Paul
Proceedings of CEISIE'09

DAVIM J. Paulo
Sustainable Manufacturing

GIORDANO Max, MATHIEU Luc, VILLENEUVE François
Product Life-Cycle Management / Geometric Variations

LUZEAUX Dominique, RUAULT Jean-René
Systems of Systems

VILLENEUVE François, MATHIEU Luc
Geometric Tolerancing of Products

2009

DIAZ Michel
Petri Nets / Fundamental Models, Verification and Applications

OZEL Tugrul, DAVIM J. Paulo
Intelligent Machining

2008

ARTIGUES Christian, DEMASSEY Sophie, NÉRON Emmanuel
Resources–Constrained Project Scheduling

BILLAUT Jean-Charles, MOUKRIM Aziz, SANLAVILLE Eric
Flexibility and Robustness in Scheduling

DOCHAIN Denis
Bioprocess Control

LOPEZ Pierre, ROUBELLAT François
Production Scheduling

THIERRY Caroline, THOMAS André, BEL Gérard
Supply Chain Simulation and Management

2007

DE LARMINAT Philippe
Analysis and Control of Linear Systems

LAMNABHI Françoise *et al.*
Taming Heterogeneity and Complexity of Embedded Control

LIMNIOS Nikolaos
Fault Trees

2006

NAJIM Kaddour
Control of Continuous Linear Systems

Printed and bound by CPI Group (UK) Ltd, Croydon, CR0 4YY

16/09/2026

14963022-0006